LOCALIZED QUALITY OF SERVICE ROUTING FOR THE INTERNET

THE KLUWER INTERNATIONAL SERIES
IN ENGINEERING AND COMPUTER SCIENCE

LOCALIZED QUALITY OF SERVICE ROUTING FOR THE INTERNET

by

Srihari Nelakuditi
University of South Carolina
U.S.A.

Zhi-Li Zhang
University of Minnesota
U.S.A.

SPRINGER SCIENCE+BUSINESS MEDIA, LLC

Library of Congress Cataloging-in-Publication

LOCALIZED QUALITY OF SERVICE ROUTING FOR THE INTERNET
by Srihari Nelakuditi, Zhi-Li Zhang
ISBN 978-1-4020-7477-6   ISBN 978-1-4615-0383-5 (eBook)
DOI 10.1007/978-1-4615-0383-5

Printed on acid-free paper.

Dedicated to our parents

Contents

List of Figures

Preface

The exponential growth of Internet brings to focus the need to control such large scale networks so that they appear as coherent, almost intelligent, organisms. It is a challenge to regulate such a complex network of heterogeneous elements with dynamically changing traffic conditions. To make such a system reliable and manageable, the decision making should be decentralized. It is desirable to find *simple local rules and strategies* that can produce *coherent and purposeful global* behavior. Furthermore, these control mechanisms must be adaptive to effectively respond to continually varying network conditions. Such adaptive, distributed, localized mechanisms would provide a scalable solution for controlling large networks. The need for such schemes arises in a variety of settings. In this monograph, we focus on *localized approach to quality of service routing*.

Routing in the current Internet focuses primarily on connectivity and typically supports only the "best-effort" datagram service. The routing protocols deployed such as OSPF use the *shortest path only* routing paradigm, where routing is optimized for a single metric such as hop count or administrative weight. While these protocols are well suited for traditional data applications such as ftp and telnet, they are not adequate for many emerging applications such as IP telephony, video on demand and teleconferencing, which require stringent delay and bandwidth guarantees. The "shortest paths" chosen for the "best effort" service may not have sufficient resources to provide the requisite service for these applications. Furthermore, with the explosive growth of the Internet traffic, the *shortest path only* routing paradigm of the current Internet also leads to unbalanced traffic distribution — links on frequently used shortest paths become increasingly congested, while links not on shortest paths are underloaded. In order to address these issues, Quality-of-Service (QoS) based routing is proposed where paths for flows are selected based upon knowledge of the resource availability at network nodes and the QoS requirements of the flows. It is expected that QoS routing will choose, among the many possible choices, a path that has sufficient resources to accommodate the QoS requirement of a given flow.

In QoS routing, some knowledge regarding the global network state is crucial in performing judicious path selection. This knowledge, for example, can be obtained through periodic information exchange among routers in a network. Under this approach, which we refer to as the *best-path* routing approach, each router constructs a global view of the network state by piecing together the QoS state information obtained from other routers, and selects the "best path" for a flow based on this global view of the network state. Best-path routing schemes work well when each source node has a reasonably *accurate* view of the network QoS state. However, as the network resource availability changes with each flow arrival and departure, maintaining an accurate network QoS state is impractical, due to the prohibitive communication and processing overheads entailed by frequent QoS state information exchange. In the presence of *inaccurate* information regarding the global network QoS state, best-path routing schemes suffer degraded performance as well as potential instability.

As a viable alternative to the best-path routing approach, we propose a novel *localized proportional routing* approach to QoS routing. Under this approach, instead of periodically exchanging information with other routers to obtain a global view of the network QoS state, a source router attempts to *infer* the network QoS state from *locally collected flow statistics* such as flow arrival/departure rates and flow blocking probabilities, and performs adaptive proportioning of flows among a set of *candidate* paths based on this local information. Under pure localized approach, the candidate path set remains static while their proportions are adjusted dynamically. A network node under localized approach can judge the quality of paths only by routing some traffic along them. So it is not possible to update the candidate path set based on local information alone. On the other hand, due to changing network conditions, a few good paths cannot be selected statically. Hence we propose a path selection procedure that dynamically selects a few good candidates based on infrequently exchanged global information. The inaccuracy in candidate path selection is cushioned by adaptively proportioning traffic among candidates. In this monograph, we first address the issue of *finding near-optimal proportions* for a given a set of candidate paths based on locally collected flow statistics. We then focus on the *selection of few good candidate paths* based on infrequently exchanged global information. The extensions to proportional routing approach for providing hierarchical routing across multiple areas in a large network is also discussed. We demonstrate that our localized adpative proportioning approach offers a scalable solution to QoS routing.

Acknowledgments

We thank Prof. David Du for his guidance throughout this project. We also wish to thank Rose Tsang for supporting us and giving us the opportunity to work with her at Sandia National Laboratories in Livermore. The collaboration with her gave a definite direction and shape to this work.

Many people have helped us at various stages of the preparation of this monograph. In particular, we like to thank Raja Harinath, Sanjai Rayadurgam, and Srivatsan Varadarajan, who were always available for discussions and gave invaluable suggestions.

Chapter 1

INTRODUCTION

Routing in the current Internet focuses primarily on connectivity and typically supports only the "best-effort" datagram service. The routing protocols deployed such as OSPF [43] use the *shortest path only* routing paradigm, where routing is optimized for a single metric such as hop count or administrative weight. While these protocols are well-suited for traditional data applications such as ftp and telnet, they are not adequate for many emerging applications such as IP telephony, video-on-demand and teleconferencing, which require stringent delay and bandwidth guarantees. The "shortest paths" chosen for the best-effort service may not have sufficient resources to provide the requisite service for these applications. Furthermore, with the explosive growth of the Internet traffic, the *shortest path only* routing paradigm of the current Internet also leads to unbalanced traffic distribution — links on frequently used shortest paths become increasingly congested, while links not on shortest paths are underloaded.

In order to address these issues, multi-path traffic engineering techniques have been proposed [5], of which Quality-of-Service (QoS) based routing [7, 11] is an important new mechanism. In QoS routing, paths for flows are selected based upon knowledge of the resource availability (referred to as *QoS state*) at network nodes (i.e., routers) and the QoS requirements of the flows. It is expected that QoS routing will select, among the many possible choices, a path that has sufficient resources to accommodate the QoS requirement of a given flow. QoS routing can significantly improve the network throughput because of its awareness of the network QoS state.

In QoS routing, some knowledge regarding the (global) network QoS state is crucial in performing judicious path selection. This knowledge, for example, can be obtained through (periodic) information exchange among routers in a network. Under this approach, which we refer to as the *global best-path* routing approach, each router constructs a global view of the network QoS state by piecing together the QoS state information obtained from other routers, and selects the "best" path for a flow based on this global view of the network state. Examples of the global best-path routing approach are various QoS routing schemes [1, 34, 66, 11, 69] based on QoS extension to the OSPF routing protocol as well as the ATM PNNI [4] routing protocol. Global best-path routing schemes work well when each source node has a reasonably *accurate* view of the network QoS state. However, as the network resource availability changes with each flow arrival and departure, maintaining an accurate network QoS state is impractical, due to the prohibitive communication and processing overheads entailed by frequent QoS state information exchange. In the presence of *inaccurate* information regarding the global network QoS state, best-path routing schemes may suffer degraded performance as well as potential instability [59, 46, 47].

As a viable alternative to the global best-path routing approach, we propose a novel *localized proportional routing* approach to QoS routing. Under this localized proportional routing approach, instead of (periodically) exchanging information with other routers to obtain a global view of the network QoS state, a source router attempts to *infer* the network QoS state from *locally collected flow statistics* such as flow arrival/departure rates and flow blocking probabilities, and performs adaptive proportioning of flows among a set of *candidate* paths based on this local information. As a result, the localized proportional routing approach avoids the drawbacks of the conventional best-path routing approach such as degraded performance in the presence of inaccurate routing information. Furthermore, it has several important advantages: *minimal* communication overhead, *no* processing overhead at *core* routers, and *easy* deployability.

We investigate an important and fundamental issue in the design of localized QoS routing schemes — the *granularity* of locally collected QoS state information and its impact on the convergence process of these schemes and their performance. We consider flow statistics collected at two different granularity levels: *link* level and *path* level. At the (*finer*) link level, a source node collects *both* the blocking statistics (i.e., whether a flow is blocked or not) of flows routed along a path from the source node to a destination node, *and*, in the case of a blocked flow, the identity of the link where the flow is blocked. The latter information can be gathered, for example, by attaching the identity of the link in the flow setup failure notification sent back to the source node. At the (*coarser*) path-level, a source node collects *only* the flow blocking statistics for each path between the source node and a destination node. Clearly, the path-level flow statistics are much easier to collect and maintain, but they also convey much less precise information regarding the (global) network QoS state.

We propose theoretical models based on the link-level and path-level flow statistics to study the impact of granularity. These models are developed based on the notion of *virtual capacity* of a link or a path as *perceived* by a source node. The virtual capacity of a link or a path is computed as a function of the amount of offered load and the corresponding observed blocking probability on that link or path. Through numerical investigation, we study the convergence process of virtual capacity based theoretical models and show that it is possible to design localized proportional routing schemes that converge to a stable point. We find that though granularity of information does have impact on the rate of convergence and the equilibrium blocking probability, the performance penalty due to coarser path-level information is not significant. Based on these theoretical results, we proceed to develop practical proportioning strategies that are simple and easy to implement. We propose two such strategies that require only path-level information: *equalization of blocking probabilities* (ebp) and *equalization of blocking rates* (ebr). We compare their performance with optimal proportional routing and show that *ebp* strategy yields near-optimal proportions.

The localized proportioning approach described above splits the traffic bound to a destination adaptively among a set of *candidate* paths. Two key questions that arise in proportional routing are how many candidate paths are needed and how to find these paths. Clearly, the number and the quality of the paths chosen as candidates dictate the performance of a proportional routing scheme. There are several reasons why it is desirable to minimize the number of paths used for routing. First, there is a significant overhead associated with establishing, maintaining and tearing down of paths. Second, the complexity of the scheme that distributes traffic among multiple paths increases considerably as the number of paths increases. Third, there could be a limit on the number of explicitly routed paths such as label switched paths in MPLS [57] that can be setup between a pair of nodes. Therefore it is desirable to use *as few paths as possible* while at the same time *minimize the blocking probability* in the network. Furthermore, it is not possible to provide each node with accurate information about the network state due to prohibitive communication and processing overheads. Hence it is important to devise candidate path selection schemes that *work well even with infrequent link state updates*. We propose such a scheme *widest disjoint paths* (wdp) that selects widest paths that are disjoint *w.r.t.* bottleneck links. It uses *infrequently* exchanged *global* information for selecting a few good paths based on their long term available bandwidths. The traffic is proportioned among the candidate paths using *local* information to cushion the short term variations in their available bandwidths. This *hybrid* approach to

QoS routing adapts at different time scales to the changing network conditions.

The above discussion assumes that each router in the network is aware of the topology and the state of the whole network. This is referred to as *flat* routing and under flat routing, each router participates in link state updates and maintains detailed information about the entire network. This introduces significant burden on every router and as the size of the network grows, the overhead at each router increases tremendously. To provide a scalable solution, *hierarchical routing* is suggested as an alternative to flat routing. Under hierarchical routing, a network is divided into multiple areas. The routing within the area is flat with each router having detailed information about routers and links in that area. But the routers have only sketchy *aggregate* information about other areas. To route traffic destined for other areas, a source router may select a partial higher level path, based on the aggregate information, that gets expanded, based on the detailed information, at the ingress border router of each area along the path. Such a hierarchical routing reduces the overhead at each router by limiting the scope of link state updates and maintaining only summary information about other areas. The hierarchical routing approach while reduces the burden on a router, introduces inaccuracy in the information available for routing. Hence the performance of a hierarchical routing scheme depends heavily on how information about an area is aggregated and how it is utilized in routing across areas.

We propose an aggregation method that summarizes the state of multiple paths between two routers using a single metric. This metric in essence captures the traffic carrying capacity of multiple paths between a pair of routers. We propose two inter-area routing schemes based on this aggregate metric: *hierarchical widest disjoint paths* (hwdp) and *hierarchical widest border routers* (hwbr). The *hwdp* scheme is a hierarchical source routing scheme where a source router selects a set of higher level skeletal paths to the destination as candidates and proportions flows among them. The *hwbr* scheme is a hierarchical next-hop routing scheme that selects only the next-hop border routers which in turn select higher level paths to the destination. Both these schemes use *wdp* scheme mentioned earlier, for intra-area routing to expand the skeletal higher level paths to actual physical paths. They essentially differ in the way the network outside an area is aggregated by a border router and propagated to the interior routers. We evaluate the performance of these hierarchical proportional routing schemes and show that these schemes with only aggregate information outperform even flat global best-path routing schemes having detailed information about the network. We argue based on our results that our *hwbr* scheme due to its low overhead and high throughput, is a suitable choice for hierarchical routing across large networks.

The rest of this monograph is organized as follows. In the next chapter, we define the problem setting we consider in this monograph. Chapter 3 presents the related work. In Chapter 4 we study the theoretical models for proportional routing and propose some practical schemes in Chapter 5. The candidate path selection scheme is described in Chapter 6 and the hierarchical proportional routing schemes are presented in Chapter 7. The conclusions and future work are discussed in Chapter 8.

Chapter 2

PROBLEM SETTING

Network support for traffic with Quality of Service (QoS) guarantees usually requires a connection establishment procedure which will select a path, perform admission control and reserve resources along the selected path. The path selection procedure in turn requires link state updates to gather information about each link in the network. In the following, we state the assumptions made in this monograph about the QoS requirements, routing procedures and the update policies. Similarly the performance metrics used for evaluating the efficacy of various QoS routing are also discussed.

1. Bandwidth Guarantees

Different applications have different quality of service requirements. Some require throughput guarantees, some end-to-end delay guarantees while others require loss rate guarantees. It is the job of the network to map these application requirements to network resources such that the requested QoS can be guaranteed. Resource provisioning that ensures both guaranteed service to applications and efficient utilization of resources is a very complex task. In this monograph, we assume that application requirements are either specified or characterized by a single network resource, bandwidth. This could be the *effective bandwidth* that captures the traffic characteristics of the application. This is not too limiting an assumption since delay constraints can either be handled by translating them to corresponding bandwidth guarantees or by bounding the number of hops during path selection. Furthermore, though there may be need for supporting applications with multiple QoS parameters, it is likely that initial deployments focus simply on bandwidth based guarantees to reduce the operational complexity. In other words, in this monograph we assume that an application requests for a specific bandwidth and the network admits an application only if it can reserve the requested bandwidth to satisfy the application requirements.

2.   Explicit Routing

The current Internet employs hop-by-hop routing that selects shortest paths periodically and maintains the path information in the routing table so that routing results in a simple table look up. Under QoS routing, different flows from various source nodes to the same destination may request different QoS. Hence it is not possible to maintain a single routing table to route flows with divergent requirements. Explicit routing is suggested as an alternative to support QoS routing. Under explicit routing, a source node selects, for each flow, an explicit path to the destination. This requires an ability to setup explicit routes in the network. This can be implemented in the current Internet using a form of loose source routing. But the overhead of carrying the complete explicit route with each packet is prohibitive. However, Multi-Protocol Label Switching (MPLS) an emerging Internet Engineering Task Force (IETF) standard [57] provides such a capability. MPLS replaces the standard destination based hop-by-hop forwarding paradigm with a label swapping forwarding paradigm. It makes explicit routing practical by allowing the explicit route to be carried only at the time label switched path is set up. This monograph assumes that such an explicit routing is supported by the network.

Under source directed QoS routing, upon arrival of a flow, the source node first selects a path that is likely to satisfy the requested QoS. This path selection is performed by the source node based on its own local view of the network state that is gathered through link state updates and previous routing attempts. The source node may prune (seemingly) *infeasible* links, with fewer resources than required by the flow, in order to select a *feasible path*. When no feasible path is found, the flow is blocked due to *routing failure*. This routing failure could be due to either lack of network resources or staleness of link state. Once a path is selected for the flow, the source then sends a setup request to reserve resources at each link along the selected path. Each node along the path performs admission control to see whether sufficient resources are available on the link to support this flow. If the link can accommodate the flow, resources are reserved on that link for the flow and the setup message is forwarded to the next link along the path. If all the links along the path have sufficient resources, the setup request is accepted and the flow is admitted. The resources reserved for a flow remain with it for the entire duration of the flow. Moreover, the route is *pinned*, i.e., all the packets of the flow follow the same path for the duration of the flow, even if a "better" path is found during that time. This helps reduce the variance in the delay experienced by the packets of a flow.

When a link does not have adequate amount of available resources, the setup request is rejected and the flow is blocked. This is categorized as *setup failure*.

A setup failure is the result of the discrepancy between the actual link state and the view of the source node. The main focus of this monograph is how to reduce the inaccuracy in a source node's view without incurring excessive update overhead and how to select "good" paths in the presence of inevitable inaccuracy. Thus a flow may be blocked due to either *routing failure* or *setup failure*. The main objective of any QoS routing scheme is to minimize the overall blocking probability. When a setup request for a flow results in failure, we can make an attempt to reroute the flow by finding an alternate path. This is known as *crankback*. While it is possible that crankbacks can decrease the probability of blocking a flow, they increase the signaling overhead. Moreover, a failed setup request could be an indication of the overall load in the network. In such a case, alternate routing may consume resources along the primary paths of other source-destination pairs increasing the blocking probability for future flows between those pairs. Thus crankbacks can potentially increase the overall blocking probability in an effort to accommodate an individual flow. In this work, we assume *no crankback*, i.e., if a setup request is rejected the flow is blocked and no attempt is made to reroute the flow.

3. Link State Updates

Current Internet routing protocols, such as OSPF, distribute connectivity information throughout the network so that "shortest" paths may be selected. Path selection under QoS routing requires resource availability information apart from the connectivity information about the network. It is suggested that current routing protocols such as OSPF be extended [17, 69] to carry *QoS state* information also in their link state updates. The QoS state of a link may be captured by its instantaneous available bandwidth, i.e., the amount of bandwidth available at the time of the update and/or average available bandwidth, i.e., the amount of bandwidth available on the average since the last update. Maintaining accurate network connectivity information usually requires an insignificant amount of routing updates since network connectivity changes are infrequent. However, maintaining accurate QoS state information requires a much greater amount of link state updates because network resource availability changes with each flow arrival and departure. Since our focus is on studying the impact of QoS state update frequency on the performance of path selection schemes, we make the following simplifying assumptions. We assume that topology does not change and links do not fail during a simulation run. Also, we ignore the link propagation delay since it does not have a bearing on the outcome of this study.

4.   Performance Metrics

The goal of QoS routing is efficient utilization of resources while ensuring quality of service to each admitted flow. In our *flow based model with bandwidth guarantees*, a flow is admitted only if the requested amount of bandwidth is available along the path through which it is routed. The quality of service for an admitted flow is guaranteed by reserving the requested amount of bandwidth for the entire duration of the flow and by reclaiming it only after the flow is departed. Under this bandwidth reservation model, a flow may be rejected at the setup time but once admitted all the packets of the flow are guaranteed to receive the requested service. Hence we only need to perform flow level simulation to study the performance of various QoS routing schemes. The objective of a QoS routing scheme is then to maximize the number of admitted flows into the network or in other words minimize the blocking probability experienced by a flow arriving in the network. Thus, a measure of performance of a QoS routing scheme is the *overall flow blocking probability*. It is computed as the ratio of the number of flows blocked and the total number flows that have arrived at the network.

There are several overheads associated with a QoS routing scheme. Measuring the blocking performance alone without the corresponding overheads involved in the implementation is not sufficient to evaluate the overall performance of a routing scheme. We categorize the overheads incurred by a QoS routing scheme into *update overhead, path computation overhead* and *path management overhead*. There is a fundamental tradeoff between the amount of overhead due to link state updates and the blocking performance of path selection schemes. The higher the frequency of updates is, the lesser the inaccuracy at a source node is and the better the blocking performance of a path selection scheme is. Ideally, we would like to have lower blocking with minimal overhead. But these are conflicting objectives and the job of a QoS routing scheme is to limit the update overhead while dealing with the corresponding inaccuracy by selecting paths intelligently. In this monograph, we assume simple periodic link state updates and we measure its overhead by the *frequency of updates*.

Another overhead associated with QoS routing is the time spent in finding a suitable path. A path may be computed on-demand upon each flow arrival or a path may be chosen from a set of precomputed paths. Since identifying a suitable path may involve searching the entire graph, the cost of computation is not insignificant. The complexity of the algorithm used for finding a path is referred to as *path computation overhead*. Finally, it is desirable to minimize the number of paths used for routing. There is a significant overhead associated with establishing, maintaining and tearing down of paths. Moreover, there could be a limit on the number of explicitly routed paths such as label switched paths in MPLS [57] that can be setup between a pair of nodes. Hence the number of paths used for routing, averaged across all source-destination pairs, is taken as a measure of the *path management overhead*. Essentially the overall objective of QoS routing is to minimize the above mentioned overheads while minimizing the overall flow blocking probability.

Chapter 3

RELATED WORK

The problem of QoS routing has been addressed in many contexts and there have been several proposals for providing QoS routing. These proposals differ in where the path is chosen (source or hop-by-hop), how the network state information is gathered (global updates or local observations), what type of information is exchanged (instantaneous available bandwidth or long-term average bandwidth), which path is selected (widest or shortest), etc. A survey of various QoS routing schemes can be found in [7]. We can broadly categorize them into *global QoS routing* schemes that are based on global link state updates and *localized QoS routing* schemes that are based on local path state observations. The following sections discuss these QoS routing approaches in detail.

1.   Global QoS Routing

The majority of QoS routing schemes [1, 11, 17, 34, 61, 66, 69] proposed so far require periodic exchange of *link QoS state* information among network nodes to obtain a *global view of the network QoS state*. Based on this *current* global view of the network state, a source node dynamically determines the "best" feasible path for a flow originating from it to a destination node. We refer to this approach to QoS routing as the *global* QoS routing approach or *global best-path routing* approach. The proposed global QoS routing schemes primarily differ in their path selection criteria and the network state update triggering mechanisms.

Path selection algorithms have to deal with the fundamental trade-off in minimizing the resource usage and balancing network load. The resource usage by a flow can be minimized by selecting the shortest path which may be heavily loaded. The network load can be balanced by choosing the least loaded path which may be longer and hence consumes more resources. Several path selection algorithms have been proposed that weigh limiting hop count and balancing network load differently. They include widest-shortest path (*wsp*) [1], shortest-widest path (*swp*) [66], and shortest-distance path (*sdp*) [34]. They all attempt to select a *feasible* path. A path is considered *feasible* if its *bottleneck bandwidth* (smallest available bandwidth along the path) is greater than or equal to the requested bandwidth. The above mentioned schemes differ in the selection of a feasible path when there are many such choices as explained below.

Widest-shortest path A path with fewest number of hops among all feasible paths. If there are several shortest feasible paths, the one with the largest bottleneck bandwidth is chosen. If more than one shortest path with same bandwidth exist, one of the paths is randomly selected.

Shortest-widest path A path with largest bottleneck bandwidth among all feasible paths. If more than one widest feasible path exist, the one with the minimum hop count is chosen. If there are several such paths with the same hop count, one of them is randomly selected.

Shortest-distance path A feasible path with the shortest distance. The distance function for a path r is defined by

$$dist(r) = \sum_{j=1}^{k} \frac{1}{C_j} \qquad (3.1)$$

where C_j is the bandwidth available on a link j along path r.

The widest-shortest path favors shorter paths thus giving higher priority to limiting resource usage, while the shortest-widest path favors wider paths thus giving higher priority to distributing network load. The shortest-distance path attempts to strike a balance by using the distance function. Among these, *wsp* is the most popular and well studied algorithm for selecting the "best" feasible path and hence we use it as a representative of global best-path routing schemes in the following discussion.

Each network node under global QoS routing approach generates link state updates informing all other nodes about the current state of the links attached to it. Various update policies are possible that differ in when an update is triggered and what information is contained in it. Most of the schemes proposed so far [1, 34, 66] exchange information about currently available bandwidth while some [18, 46] exchange information regarding the probability that a requested bandwidth is available. The mechanisms used to trigger updates can be based on *threshold, class* or *timer*. In case of threshold based policy, an update is triggered whenever the relative change from the previously advertised to the current link state exceeds a certain threshold. Class based policies partition the available bandwidth into multiple classes and trigger an update whenever the current link state value crosses a class boundary. The partitioning into classes could be either fixed-size or exponentially distributed. Timer based triggers may be used to generate updates periodically. They, referred to as *clamp-down* timers, are often used in conjunction with one of the above triggers to enforce a minimum spacing between two consecutive updates. While the periodic updates are simple to implement, more complex change based triggers attempt to avoid unnecessary updates by generating an update only when the link state changes *significantly* from previously advertised state. Another trade-off in these schemes is between the frequency of updates and the accuracy of network state information available at each node for path selection.

Current Internet routing protocols, such as OSPF, distribute connectivity information throughout the network so that "shortest" paths may be selected. Maintaining accurate network connectivity information usually requires an insignificant amount of routing updates since network connectivity changes are infrequent. However, maintaining *accurate* network QoS state requires *frequent* information exchanges among the network nodes because network resource availability changes with each flow arrival and departure. The prohibitive communication and processing overheads entailed by such frequent QoS state updates preclude the possibility of *always* providing each node with an accurate view of the current network QoS state. Consequently, *the network QoS state information acquired at a source node can quickly become out-of-date when the QoS state update interval is large relative to the flow dynamics.* Under these circumstances, exchanging QoS state information among network nodes is superfluous. Furthermore, path selection based on a *deterministic* algorithm such as Dijkstra's shortest path algorithm, where *stale QoS state information is treated as accurate*, does not seem to be judicious. The *best* path selection based on *inaccurate* information could cause instability: after one QoS state update, many source nodes choose paths with shared links because of their perceived available bandwidth, therefore causing over-utilization of these links. After the next QoS state update, the source nodes would avoid the paths with these shared links, resulting in their under-utilization. This oscillating behavior can have severe impact on the system performance, when the QoS state update interval is large. Due to these drawbacks, it has been shown that when the QoS update interval is large relative to the flow dynamics, the performance of global QoS routing schemes degrades drastically [46, 47].

Several solutions have been proposed to deal with inaccuracy that is inevitable in the information available to path selection process. One approach [1] categorizes the inaccuracy into *systematic* and *random* based on the type of update triggers used. When a change based trigger is employed it is possible to infer the range for actual link metric value given its last advertised value. This type of systematic inaccuracy could be suitably accounted for by a path selection algorithm and thereby choose a path that is most likely to have the required resources. In [18] a path selection algorithm is proposed to find the *most reliable path* assuming that information regarding the probability $p_l(x)$ that a link l can accommodate a flow which requires x units of bandwidth is known to the source node. This is further experimented in their paper [2] on *safety-based routing*, where safety, i.e., $p_l(x)$, of a link l for a bandwidth x is inferred from its last advertised available bandwidth value assuming that a change based update triggering policy is used. However when large *clamp-down* timers are used, it is almost impossible to estimate the amount of inaccuracy. Some randomization algorithms are proposed [2] to cope with this kind of random inaccuracy where advertised bandwidth values are used only as clues in selecting paths. We have proposed a probabilistic approach [46] where network nodes exchange link availability probability information similar to $p_l(x)$. But instead of choosing the *best* or *safest* path our algorithm distributes load across the network in accordance with availability of links while favoring shorter paths.

The work more relevant to ours is the distributed routing scheme proposed in [8] where a set of multiple paths are probed in parallel, using tickets, for a satisfactory path. However, this approach requires the distribution and processing of these tickets by intermediate nodes. Minimum interference routing [30] is a scheme proposed recently that selects a path that interferes least with the routing of future flows. While this scheme provides good routing performance, it has significant computational overhead. The proportional routing approach presented in this monograph achieves the similar effect by gradually adapting the flow proportions assigned to paths based on their blocking probabilities which is an indirect measure of interference of paths.

While all the above remedial schemes reduce the impact of inaccuracy on the performance of path selection, they either work well only in some cases or introduce additional overhead at core routers. For easy deployability and scalability of QoS routing, we need to devise schemes that perform well without introducing more complexity at core routers and more communication overhead on network than the current routing protocols. As an alternative to global QoS routing, *localized* QoS routing approach is proposed, where *no global QoS state information exchange among network nodes is needed*. Instead, source nodes infer the network QoS state based on flow blocking statistics collected *locally*, and perform flow routing using this *localized* view of the network QoS state. Some of the *localized* QoS routing schemes are described in the following sections.

2.   Localized QoS Routing

A source node under QoS routing, upon receiving a request with specific QoS requirements, selects a suitable path to the destination node based on its view of resource availability. It then sends a connection request to reserve resources at each node along the path. It is possible that sufficient resources are not available along the chosen path either because of stale routing information at the source node or because of changes in the network state while the connection is being established. In such a case, the request is rejected and the flow is blocked. Localized QoS routing approach attempts to infer the network state from these flow blocking statistics and performs path selection based on this local information. Several localized dynamic routing schemes have been proposed in the context of telephone networks. Here we consider two such schemes based on sticky routing and learning automata that make use of the feedback information regarding flow admission or rejection for routing future flows.

2.1 Sticky Random Routing

The *dynamic alternative routing* (dar) is a well known routing scheme [15] where a source always tries the direct one-link path to the destination first and in case of a crankback chooses a two-link path using *sticky random routing* (srr). Since in our setting we do not consider re-routing, the *srr* scheme (equivalent to *dar* with a dummy direct link) is used for comparison. The *srr* scheme remembers a path known as *preferred* path for each destination. A flow to a destination is always routed through its corresponding preferred path. If the connection setup is successful, the preferred path remains same. But in case of a failure, the flow is blocked and a new preferred path is chosen randomly from set of feasible paths to that destination excluding the current preferred path. The *srr* scheme essentially sticks to a path as long as it can accommodate offered traffic.

The analysis of *dar* presented in [15] shows that *dar* equalizes the blocking rates over two-link paths for each source destination pair. It claims that overflow streams, i.e., flows directed to two-link paths, under *dar* can be modeled as if they arise from proportional routing, with proportions depending on the blocking rates of links. But it also cautions that the approximation procedure used in the analysis could break down if the overflow is large and needs to be spread over a number of alternatives. This is precisely the case with networks like Internet that may have more than one minhop path and many alternative paths between each source-destination pair.

2.2 Learning Automata based Routing

An application of automata to the routing problem is given by Narendra and Mars [44]. The incoming flows are offered to a path r according to a probability distribution p_r, which is updated using feedback information regarding flow admission or rejection. These schemes reward a path on which a flow is successful and punish a path on which a flow fails. If a route i is chosen at time n and the flow is successful, then updating is

$$p_i(n+1) = p_i(n) + a(1 - p_i(n))$$

$$p_j(n+1) = (1-a)p_j(n) \quad j \neq i$$

while if the flow fails

$$p_i(n+1) = (1 - \epsilon)p_i(n))$$

$$p_j(n+1) = \frac{\epsilon}{r-1} + (1-\epsilon)p_j(n) \quad j \neq i$$

where a and ϵ are adjustable parameters, $0 < a < 1$, $0 < \epsilon < 1$ with ϵ small compared with a, and a is itself usually small, so that the updating is gradual. Under certain assumptions [44, 45, 62] show that $L_{R-\epsilon P}$ automata tends to approximately equalize blocking probabilities, b_r, while L_{R-P} automata for which $\epsilon = a$ in the above equalizes blocking rates $(p_r b_r)$. One problem with this scheme is that probability associated with a path is changed per every flow arrival. In addition, no account is taken of the length of a path and also the selection of candidate paths. In the later chapters we compare the performance of these schemes with our schemes and show that our schemes yield much lower blocking probability.

3.   Hybrid QoS Routing

As a viable alternative to the best-path routing approach, we proposed [47, 48] a novel *proportional routing* approach to QoS routing. Under this proportional routing approach, a source router uses *locally collected flow statistics* such as flow arrival/departure rates and flow blocking probabilities, and performs adaptive proportioning of flows among a set of *candidate* paths based on this local information. This localized proportional routing approach is somewhat similar in spirit to the dynamic routing schemes in telephone networks [29] described above. However, the actual mechanisms for adaptive proportioning are quite different. Moreover, we enhance the performance of localized proportional routing by selecting a few good candidate paths.

Under pure localized approach, the candidate path set remains static while their proportions are adjusted dynamically. A network node under localized approach can judge the quality of paths only by routing some traffic along them. So it is not possible to update the candidate path set based on local information alone. On the other hand, due to changing network conditions, a few good candidate paths cannot be selected statically. Hence we propose [50, 51] a *hybrid* approach to proportional routing, where a few good candidate paths are selected dynamically based on *infrequently* exchanged global information. The resulting inaccuracy is cushioned by adaptively proportioning traffic among multiple "good" candidate paths instead of routing all the traffic along the "best" path. In the following chapters, we first address the question of how to proportion traffic adaptively among a set of candidate paths using only local information. We then turn to the issue of how to select a few good candidate paths based on infrequently exchanged global information.

Chapter 4

LOCALIZED PROPORTIONAL ROUTING: THEORETICAL MODELS

In this chapter we study several issues in the design of *localized* QoS routing schemes, which make *local* routing decisions based on *locally collected* QoS state information. In particular, we investigate the granularity of local QoS state information and its impact on the design of localized QoS routing schemes from a theoretical perspective. We develop two theoretical models for studying localized proportional routing: one using the locally collected link-level QoS state information, and the other using locally collected path-level QoS state information. We compare the performance of these localized proportional routing models with that of a global optimal proportional model that has knowledge of the global network QoS state. We first describe the global optimal proportioning scheme and then localized adaptive proportioning schemes.

1. Global Optimal Proportional Routing

As a basis for comparing the performance of localized proportional QoS routing models, in this section we present the global optimal proportional routing model, which has been studied extensively in the literature (see [58] and references therein). In this model, we assume that each source node knows the *complete topology information of the network (including the capacity of each link)* as well as the *offered traffic load between every source-destination pair*. With the global knowledge of the network topology and offered traffic loads, the *optimal proportions*, for distributing flows among the paths between each source-destination pair, can be computed as described below.

Consider an arbitrary network topology with N nodes and L links (see Figure 4.1 for example topologies). For $l = 1, 2, \ldots, L$, the capacity of link l is $c_l > 0$, which is assumed to be fixed and known. The links are unidirectional, i.e., carry traffic in one direction only. Let $\sigma = (s, d)$ denote a source-destination pair in the network. Let λ_σ denote the average arrival rate of flows

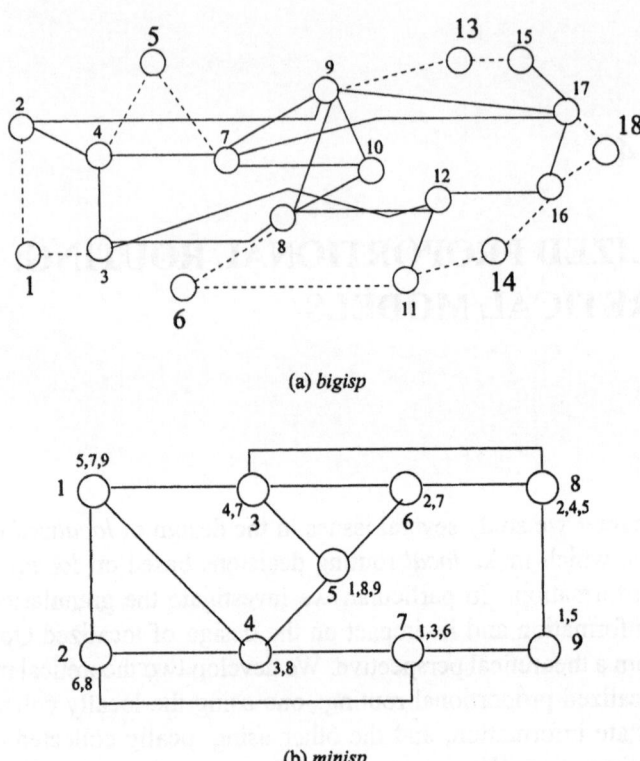

(a) *bigisp*

(b) *minisp*

Figure 4.1.   The topologies used in our study

arriving at source node s destined for node d. The average holding time of the flows is μ_σ. Recall that each flow is assumed to request one unit of bandwidth, and that the flow arrivals are Poisson, and flow holding times are exponentially distributed. Thus the offered load between the source-destination pair σ is $\nu_\sigma = \lambda_\sigma / \mu_\sigma$. Let R_σ denote the set of (explicit-routed) paths for routing flows between the source-destination pair σ. The global optimal proportional problem can be formulated [24, 25, 28] as the problem of finding the optimal proportions $\{\alpha_r^*, r \in R_\sigma\}$ for each source-destination pair σ and its route set R_σ, where $\sum_{r \in R_\sigma} \alpha_r^* = 1$, such that the overall flow blocking probability in the network is minimized. Or equivalently, finding the optimal proportions $\{\alpha_r^*, r \in R_\sigma\}$ such that the total carried traffic in the network,

$$W = \sum_\sigma \sum_{r \in R_\sigma} \alpha_r \nu_\sigma (1 - b_r) \qquad (4.1)$$

is maximized.

In (4.1), b_r denotes the blocking probability along path r. Under the *link independence* assumption [24, 25, 28], b_r can be expressed as follows:

$$b_r = 1 - \prod_{l \in r}(1 - b_l) \qquad (4.2)$$

where $l \in r$ means that link l is part of route r, and b_l is the blocking probability of link l. The blocking probability b_l of link l is in turn given by the Erlang loss formula,

$$b_l = E(\nu_l, c_l) = \frac{\frac{\nu_l^{c_l}}{c_l!}}{\sum_{n=0}^{c_l} \frac{\nu_l^n}{n!}}. \qquad (4.3)$$

Here the load offered on link l, ν_l, is the sum of all the *reduced* loads (i.e., after *independent load thinning*) from any source-destination pair σ which has a route passing through link l. Namely,

$$\nu_l = \sum_{\sigma} \sum_{r \in R_\sigma : l \in r} \alpha_r^* \nu_\sigma \prod_{m \in r - \{l\}} (1 - b_m) \qquad (4.4)$$

The global optimal proportional routing problem (4.1) is a *constrained non-linear optimization problem* and can be solved using an iterative procedure based on the *Sequential Quadratic Programming (SQP) method* [54, 10]. Each stage of the iterative procedure has two steps. First, given a set of flow proportions α_r, the fixed-point equations (4.3) and (4.4) involving b_l's and ν_l's are solved. Using these values, W given by equation 4.1 is recomputed. Then this algorithm essentially searches for a new set of improved flow proportions based on the revenue W. The complete process needs to be repeated for various initial conditions to search for the potential global optimal solution. The optimization procedure (referred to as *optpropo*) used in our study is given in Figure 4.3. Figure 4.2 summarizes the notation used in this procedure (and the rest of the monograph).

The global optimal proportional routing model presented above assumes that the bandwidth requests of flows are homogeneous (i.e., one unit of bandwidth). In the case that flows have heterogeneous bandwidth requests, the Erlang loss formula can be extended [23, 55] and the global optimal proportional routing problem can be analogously formulated and solved using multi-rate loss models [40, 58]. The computational complexity of the exact solution is generally prohibitive for large networks. To address this problem, several approximation algorithms have also been proposed [40, 42]. In particular, when the capacity of a link is large, the blocking probability of a flow of type i can be approximated as follows [56]. Suppose that type i flow requests for d_i units of bandwidth and the load of type i flows on link l is ν_l^i. The blocking probability for type i flows on link l is given by $b_l^i = \frac{d_i}{\delta} E(\frac{\sum \nu_l^i d_i}{\delta}, \frac{c_l}{\delta})$, where δ is an

$N:$	$\{1, 2, ..n\}$, vector of n nodes in the topology of the given network.
$L:$	$\{1, 2, ..l\}$, vector of l links in the topology of the given network.
$C:$	$\{c_1, c_2, ..c_l\}$, vector of capacities of the links, $c_i > 0$.
$\sigma:$	a pair (s, d), where $s, d \in N$.
$r:$	a path, i.e a set of links $\in L$, from source to destination.
$R_\sigma:$	a set of paths between a source-destination pair σ.
$\nu_l:$	load on the link $l \in L$.
$\nu_\sigma:$	load on the source-destination pair σ.
$\nu:$	$n \times n$ traffic matrix, i.e, $[\nu_\sigma]_{\forall \sigma \in N \times N}$.
$b_l:$	blocking probability of link $l \in L$.
$b_r:$	blocking probability of path r.
$\alpha_r:$	load proportion along $r \in R_\sigma$ for a particular σ.

Figure 4.2.   Notation

```
1.    PROCEDURE optpropo(N, L, C, R, ν)
2.        Start with some arbitrary proportions αr^(0), ∀r ∈ Rσ, ∀σ = (s, d), s, d ∈ N.
3.        Use Sequential Quadratic Programming (SQP) method for solving (in i^th step):
4.            maximize W = Σσ Σr∈Rσ αr^(i) νσ (1 − br^(i))
5.            under the constraints Σr∈Rσ αr^(i) = 1.
6.            where br^(i) = 1 − ∏l∈r (1 − bl^(i))
7.        Use Gauss-Newton method to solve the following nonlinear equations for bl^(i)
8.            bl^(i) = E(νl^(i), cl)
9.            νl^(i) = Σσ Σr∈Rσ:l∈r αr^(i) νσ ∏m∈r−{l} (1 − bm^(i))
10.   END PROCEDURE
```

Figure 4.3.   The *optpropo* procedure

"equivalent rate" given by $\delta = \frac{\sum \nu_i^i d_i^2}{\sum \nu_i^i d_i}$. In other words, the ratio of blocking probabilities of flow types i and j would be same as the ratio of their bandwidth requests, i.e., $\frac{b_i}{b_j} \approx \frac{d_i}{d_j}$. We later argue that in this realistic scenario of link capacity being much larger than an individual flow's bandwidth request, it is not necessary to distinguish between different types of flows for the purpose of proportional routing.

2. Localized Proportional Routing

We now turn our attention to the problem of modeling *localized* proportional routing. Unlike in the global case, in localized proportional routing we assume that each source node has only a *local* (and thus *partial*) view of the network state. For example, a source node may only have knowledge of the offered traffic loads between the source-destination pairs originating from itself. Also, it may have partial network topology information only (in particular, the link capacity information may not be available to a source node). As mentioned in the introduction, in this chapter we will focus on local QoS state information gathered at two different granularity levels: the *link* level and the *path* level. At the (finer) link level, each source node can collect the following information locally: 1) the offered traffic load of flows from the source to a destination; 2) the number of flows routed along a path from the source to a destination that are blocked; and 3) in the case where a flow is blocked, the identity of the link at which the flow is blocked. The third type of information can be made available to a source node by simply piggybacking the identity of the link at which a flow is blocked in the flow setup failure notification sent back to the source node. At the (coarser) path level we assume that each source node only collects the first and second types of the local information listed above. In other words, when a flow is blocked, the source node does not have the knowledge of the link identity at which the flow is blocked. As a result, the path-level local information provides a source node with a much "vaguer" view of the global network QoS state.

2.1 Virtual Capacity Model

Given only locally collected flow statistics, determining "optimal" proportions for distributing flows among multiple paths between a source-destination pair becomes a difficult problem. In particular, since each source node does *not* know the capacity of a link and the total offered load on the link, the Erlang loss formula cannot be directly used to derive flow blocking probability at a link. To address this problem, we introduce the notion of *virtual capacity* of a link (or a path) *perceived by a source node*. For a link l, let $\nu_{s,l}$ be the load placed by a source node s, and $b_{s,l}$ be the blocking probability observed by node s. Intuitively, the virtual capacity, $vc_{s,l}$, of link l perceived by the source node s is the (perceived) amount of bandwidth consumed by the flows routed from source s along link l, given the observed blocking probability $b_{s,l}$. Formally, $vc_{s,l}$ is defined via the inverse of the Erlang loss formula as follows:

$$vc_{s,l} = E^{-1}(b_{s,l}, \nu_{s,l}), \qquad (4.5)$$

where $E^{-1}(b, \nu) := \min\{c : E(\nu_{s,l}, c) \leq b_{s,l}\}$, the inverse function of the Erlang loss formula with respect to the capacity. The virtual capacity of a *path* can be defined analogously by replacing the link l with a path r, $\nu_{l,s}$ and $b_{l,s}$ with $\nu_{r,s}$ and $b_{r,s}$, the load offered and blocking probability observed by node s along path r.

Note that $E^{-1}(b_{s,l}, \nu_{s,l})$ defined above is an integer-valued function. This means that if we vary either the blocking probability $b_{s,l}$ or the offered load $\nu_{s,l}$ slightly, $E^{-1}(b_{s,l}, \nu_{s,l})$ is likely to yield exactly the same virtual capacity value. This is an undesirable property that could cause some potential problem in the convergence process of localized proportional routing schemes we study later. To circumvent this problem, we resort to the *continuous* version of the

Erlang loss formula defined in [13]:

$$\tilde{E}(\nu, c) = \frac{1}{\int_0^\infty (1 + \frac{x}{\nu})^c e^{-x} dx} \tag{4.6}$$

It can be shown that the continuous version of the Erlang loss formula $\tilde{E}(\nu, c)$ defined above is analytic in c, and coincides with the discrete version $E(\nu, c)$ when c is an integer. Therefore, its inverse function $\tilde{E}^{-1}(b, \nu)$ with respect to c is well-defined. (It is easy to see that the inverse of the Erlang loss formula with respect to the offered load is also well-defined. This inverse will also be used in our localized proportional QoS routing schemes, as will be seen later.) In general, computing $\tilde{E}(\nu, c)$ and its inverses using (4.6) directly are quite difficult and time-consuming. Approximation methods for computing $\tilde{E}(\nu, c)$ and its inverse have been proposed in [13, 22]. In the rest of the chapter, the continuous version of the Erlang loss formula will be used, and for simplicity of notation, we will drop the superscript ˜.

The notion of virtual capacity defined above has several interesting and important properties that are key to our study of localized (adaptive) proportional QoS routing. First of all, it is clear that the virtual capacity of a link or path can be computed solely based on local information (e.g., load offered and blocking probability observed by a source node). Second, the notion of virtual capacity provides a *quantitative* measure of capacity share of a link or path grabbed by the flows originated from a source node. The larger the load a source node offers on a link or a path, the more capacity share the node grabs. To see this, consider a link of capacity c_l, which is shared by traffic originated from source nodes $s_1, \ldots, s_n$. Suppose the offered load from each node s_i is ν_{l,s_i}. Then the blocking probability b_l on link l, as is observed by each node s_i, is given by

$$b_l = E\left(\sum_{i=1}^n \nu_{l,s_i}, c_l\right) \tag{4.7}$$

Therefore, the virtual capacity perceived by node s_i is

$$vc_{l,s_i} = E^{-1}(b_l, \nu_{l,s_i}) \tag{4.8}$$

Clearly, the large ν_{l,s_i} is, the bigger vc_{l,s_i} is, as the observed blocking probability by each node s_i, $b_{l,s_i} = b_l$, is determined by the *total* offered load and the capacity of the link, not the individual load offered by each source. Third, the virtual capacity perceived by a node is a function of *both* its offered load *and* the observed blocking probability, which changes as the *overall* load on a link or a path varies. Consequently, a node can adjust its offered load to effect a change in the observed blocking probability, or as a response to the change in the observed blocking probability. The notion of virtual capacity therefore provides a theoretical basis for the analysis of how flow proportions should be adjusted based on locally collected statistics. We can extend this notion of virtual capacity for heterogeneous traffic also and compute class based proportions by maintaining statistics separately for each traffic class.

Based on the notion of virtual capacity, in the next section we develop two theoretical models for localized proportional routing: *virtual link based minimization (vlm)* and *virtual path based minimization (vpm)*, that compute flow proportions using, respectively, link-level and path-level flow statistics collected locally at source nodes. In both models, each source collects local QoS state information, and based on this local QoS state information, periodically recomputes flow proportions assigned to the paths from the source to a destination. This distributed dynamic adaptation procedure can be viewed as an iterative process where in each iteration each source independently attempts to minimize the observed blocking probability by adjusting the amount of traffic routed through each path. These models differ in the type and granularity of the local QoS state information collected, and therefore in the manner that the flow proportions are derived.

2.2   Virtual Link based Minimization

In the virtual link based minimization model, a source collects link-level flow blocking statistics with the assistance from the connection admission control (CAC) module. We assume that whenever a flow setup request fails at a link, the identity of that link is also recorded and piggybacked to the source. The CAC module at the source node informs the QoS routing module of the flow setup failure and the identity of the link where the flow is blocked. Such link-level flow blocking information can be gathered by a source with very little overhead on the network.

With the locally collected link-level flow statistics, a source knows exactly the offered traffic load on a link contributed by flows originating from that source. Unlike the global routing model, the source, however, does not have any information regarding the traffic loads offered by the other sources on the link. It neither has any knowledge of the capacity of the link. The source can only infer the state of the link from the flow blocking probability at the link it observes. Using the notion of virtual capacity of a link, the source can infer its share of the bandwidth at each link, and piece together a partial *virtual view* of the network from its own perspective.

This notion of virtual network view of a source can be illustrated using Figure 4.4. First suppose that node 2 is the only source in the network and nodes 6 and 8 are its destinations. There are two minhop paths $2 \rightarrow 1 \rightarrow 3 \rightarrow 6$ and $2 \rightarrow 4 \rightarrow 5 \rightarrow 6$ to node 6. Similarly $2 \rightarrow 1 \rightarrow 3 \rightarrow 8$ and $2 \rightarrow 7 \rightarrow 9 \rightarrow 8$ are the minhop paths to node 8. All the links along these paths form the virtual network view of source 2. When node 2 is the only source and links are not shared by other sources, the virtual capacities of these links would

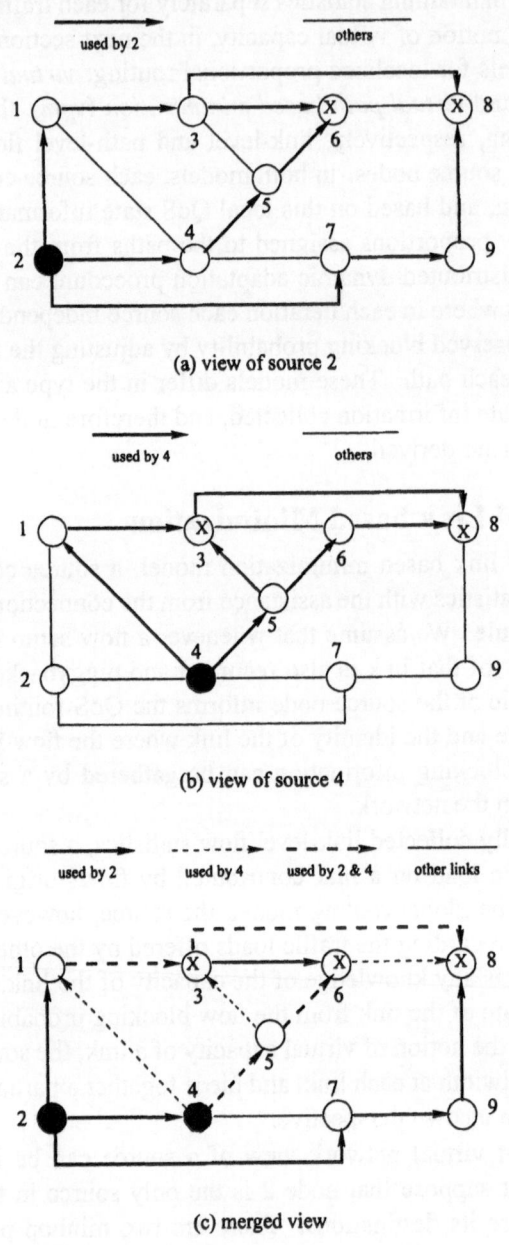

Figure 4.4.   Virtual network views of sources 2 and 4

1.	PROCEDURE VLM(s)
2.	For each link $l \in L$
3.	Compute virtual capacity $vc_{s,l}^{(n)} = E^{-1}(\nu_{s,l}^{(n)}, b_{s,l}^{(n)})$
4.	For each path $r \in R_s$
5.	Assign new load $\nu_r^{(n+1)}$ such that
6.	$\sum_{r \in R_s} \nu_r^{(n+1)}(1 - b_r)$ is maximum, where
7.	$b_r = 1 - \prod_{l \in r}(1 - b_l)$
8.	$b_l = E(\nu_{s,l}^{(n+1)}, vc_{s,l}^{(n)})$
9.	$\nu_{s,l}^{(n+1)} = \sum_{r \in R_s : l \in r} \nu_r^{(n+1)} \prod_{m \in r - \{l\}}(1 - b_m)$
10.	$\sum_{r \in R_\sigma} \nu_r^{(n+1)} = \nu_\sigma, \forall \sigma$
11.	END PROCEDURE

Figure 4.5. The *vlm* procedure at source node s

be the same as the actual capacities. Thus the network in Figure 4.4(a) with only the thick links correspond to the virtual network of source 2. Similarly, the virtual network view of source 4 is shown in Figure 4.4(b), where nodes 3 and 8 are its destinations. The thick links correspond to two minhop paths to node 3 and three minhop paths to node 8. Once again when node 4 is the only source in the network, the thick links form the virtual network of source 4.

Now consider the case where both nodes 2 and 4 are the sources. The merged virtual view of these sources is shown in Figure 4.4(c). Here the thick links are used by only source 2 and the thin dashed links are used by source 4 only. Both sources use the thick dashed links $1 \rightarrow 3$, $3 \rightarrow 8$, $4 \rightarrow 5$, and $5 \rightarrow 6$ to route their traffic. Under the link-level localized QoS routing model, each source does not have any knowledge about this sharing. This sharing is indirectly reflected in the flow blocking probability on the links observed by each source, which leads each source to derive its share of the link capacity using the link virtual capacity.

With the virtual network view, each source can employ a localized version of the global optimal proportional routing scheme (*optpropo*) to compute the "optimal" flow proportions for each of its destinations: we replace the actual capacity of a link by its virtual link capacity, and only offered traffic loads from the source are used to compute the optimal flow proportions for the source. The resulting optimization procedure, referred to as the virtual link based minimization (*vlm*) procedure, is shown in Figure 4.5, where s is a source node. This localized flow proportioning scheme is an iterative process where each iteration is performed after an observation interval by each source asynchronously. In the nth iteration, the current virtual capacity $vc_{s,l}^{(n)}$ of each link l with respect to s, is computed, based on the current offered load $\nu_{s,l}^{(n)}$ and the corresponding observed blocking probability $b_{s,l}^{(n)}$ (lines 2-3). The local

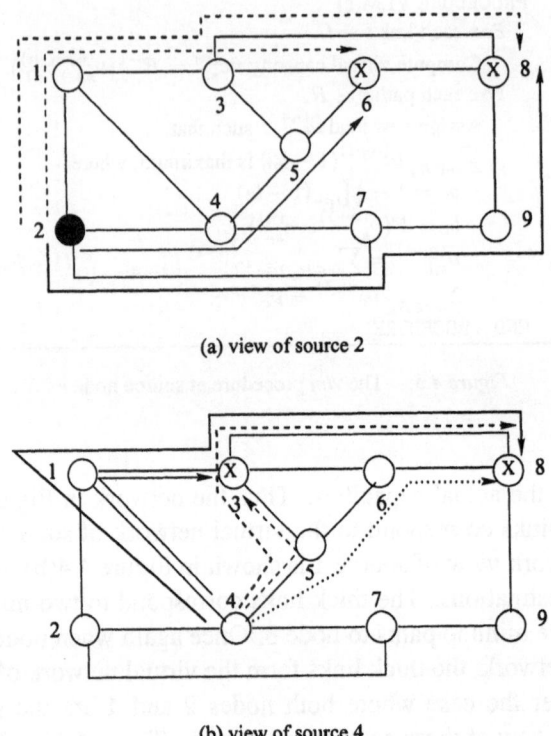

(a) view of source 2

(b) view of source 4

Figure 4.6.   Virtual network views of sources 2 and 4

minimization is then performed on the virtual network thus formed with each link l having the capacity $vc_{s,l}^{(n)}$ (lines 4-10).

2.3   Virtual Path based Minimization

In the virtual path based minimization (*vpm*) model each source collects only path-level flow statistics: the number of flows routed along each path between the source to a destination, and the number of flows blocked along the path. Unlike the link-level localized QoS routing model, here we assume that the identity of the link at which a flow is blocked is *not* available to a source.

With only locally collected path-level flow statistics, a source does not have any ways to infer the QoS state of any individual link. A source can only obtain some knowledge about the "quality" of a path based on the traffic offered on the path and the corresponding observed flow blocking probability along the path. Similar to the virtual link based QoS routing model, in the virtual path based routing model we associate a virtual network with each source-destination pair, using the notion of virtual capacity of a path. Consider a

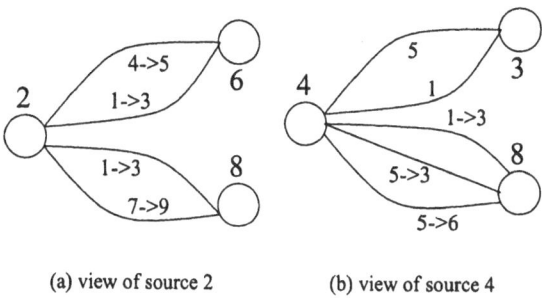

(a) view of source 2   (b) view of source 4

Figure 4.7.   Virtual network views of sources 2 and 4

source-destination pair (s, d). Suppose there are k (explicit-routed) paths between source s and destination d. Using the notion of virtual capacity of a path, we treat these k paths *as if they were disjoint and each consisted of a single virtual link*. The virtual capacity of a path r is represented by vc_r, which is determined by the offered load from source s to destination d along route r and the observed blocking probability b_r of flows routed along the route. Although the real network topology of these paths may be very complex (e.g., multiple paths have shared links, or share links with other source-destination pairs), the notion of virtual capacity of a path allows us to circumvent these difficulties by essentially capturing the "capacity share" of flows routed along various paths.

The concept of virtual capacity of a path can be illustrated using the figure shown in Figure 4.6. The virtual path view of a source can be understood as follows. Imagine a network where each physical link is split into multiple virtual links, one per each path passing through that link. In such a network all paths would be mutually disjoint. For example, Figure 4.7(a) represents virtual path view of source 2 corresponding to the physical network in Figure 4.6(a). Similarly, Figure 4.7(b) corresponds to that of Figure 4.6(b). Here the link $1 \rightarrow 3$ is shared by paths $2 \rightarrow 1 \rightarrow 3 \rightarrow 6$, $2 \rightarrow 1 \rightarrow 3 \rightarrow 8$, $4 \rightarrow 1 \rightarrow 3$ and $4 \rightarrow 1 \rightarrow 3 \rightarrow 8$. These paths can be made to appear disjoint by splitting it into 4 virtual links.

Given the path-level virtual network view for a source-destination pair, the "optimal" flow proportions for the paths between the pair can be computed to minimize the overall flow blocking probability experienced by the flows routed along these paths. Formally, consider a source-destination pair σ. Let R_σ denote the set of paths between the source-destination pair σ. For each path $r \in R_\sigma$ let vc_r denotes the virtual capacity of the path (perceived by the source-destination pair σ). The flow proportions for the paths can be computed using an iterative procedure, referred to as the virtual path based minimization (*vpm*) procedure, as is shown in Figure 4.8. In this procedure, the virtual ca-

1.	PROCEDURE VPM(σ)
2.	For each path $r \in R_\sigma$
3.	Compute virtual capacity $vc_r^{(n)} = E^{-1}(\nu_r^{(n)}, b_r^{(n)})$
4.	For each path $r \in R_\sigma$
5.	Find new load $\nu_r^{(n+1)}$ such that
6.	$\sum_{r \in R_\sigma} \nu_r^{(n+1)} E(vc_r^{(n)}, \nu_r^{(n+1)})$ is minimum
7.	$\sum_{r \in R_\sigma} \nu_r^{(n+1)} = \nu_\sigma$
8.	END PROCEDURE

Figure 4.8.   The *vpm* procedure for a pair σ

pacity $vc_r^{(n)}$ of each path r is computed using the Erlang inverse formula, given the current offered load $\nu_r^{(n)}$ along the path r and the corresponding observed blocking probability $b_r^{(n)}$. Based on these path virtual capacities, new loads $\{\nu_r^{(n+1)}\}$ are reassigned to paths such that $\sum_{r \in R_\sigma} \nu_r^{(n+1)} E(vc_r^{(n)}, \nu_r^{(n+1)})$ is minimized. This procedure is performed iteratively and independently at each source for all the source-destination pairs originating at the source.

Before we leave this section, it is interesting to contrast the link based and path based localized proportional routing model with the global optimal proportional routing model in the way they handle the sharing of links among paths. While the global model is aware of how the links are shared by all the paths between any source to any destination, the localized link-level model is only aware of sharing of links among the paths from the same source. The localized path-level model is completely oblivious of any link sharing. However, this lack of knowledge about explicit sharing between paths is somewhat compensated by the notion of virtual capacity, which indirectly accounts for the effect of link sharing. Moreover, the localized models make up for the absence of such knowledge by employing an iterative process to compute flow proportions, in an attempt to approach the optimal flow proportioning. This iterative procedure can be thought of as continually refining the (partial) virtual network view of each source. Each source uses its virtual network view to compute flow proportions for routing flows along various paths, and this in turn improves the virtual network views of all the sources. This iterative procedure eventually converges to an equilibrium state that yields near-optimal flow proportions.

2.4   Performance Comparison

In this section, we demonstrate the convergence process of the localized proportional routing models, and compare their stable performance with that of the global optimal proportional routing model through numerical investigation. Before we present the results, we first describe the system setup.

System Setup

The topologies used in our study are shown in Figure 4.1. The *bigisp* is the topology of an ISP backbone network used in [1, 34] also. There are two types of links: *solid* and *dotted*. All solid links have same capacity with C_1 units of bandwidth and similarly all the dotted links have C_2 units. For simplicity, all the links are assumed to be bidirectional and of equal capacity in each direction. The nodes labeled with a bigger font are considered to be source (ingress) or destination (egress) nodes. The *minisp* topology is almost like the core of the *bigisp* topology. It has only solid links and for each source a subset of nodes (shown in smaller font) are chosen as destination nodes. By default, results presented in this section correspond to *minisp* topology and we explicitly mention whenever *bigisp* is used.

The flow dynamics of the network are modeled as follows (similar to the model used in [59]). Each flow is assumed to require one unit of bandwidth. Flows arrive at a source node according to a Poisson process with rate λ. The incoming traffic at a source is uniformly split among its destination nodes. The holding time of a flow is exponentially distributed with mean $1/\mu$. Following [59], the offered network load is given by $\rho = \lambda N \bar{h} / \mu (L_1 C_1 + L_2 C_2)$, where N is the number of source nodes, L_1 and L_2 are the number of solid and dotted links respectively, and $\bar{h}$ is the mean number of hops per flow, averaged across all source-destination pairs. The parameters used in our study are $C_1 = 20$, $C_2 = 30$, $\mu = 1$ minute. The topology specific parameters for *minisp* are $N = 9$, $L_1 = 26$, $L_2 = 0$, $\bar{h} = 2.64$. Similarly for *bigisp* these parameters are $N = 6$, $L_1 = 36$, $L_2 = 24$, $\bar{h} = 3.27$. The average arrival rate at a source node λ is set depending upon the desired load.

Convergence

Each source under the localized proportional routing schemes observes either link-level or path-level flow blocking probabilities and periodically recomputes flow proportions for routing flows among its paths. These reassignments of flow proportions are done independently and asynchronously by each source in a distributed manner. However with the help of the virtual capacity model, they indirectly cooperate with each other and gradually converge to near-optimal proportions as demonstrated below.

Figure 4.9(a) shows the convergence process of the localized routing schemes. The load ρ is set to 0.50 on *minisp*. Only the minhop paths are chosen as the candidate paths for each source-destination pair. The average period between recomputations is set to 1. The overall blocking probability is plotted as a function of time (i.e., the number of iterations). The performance of the global optimal routing scheme is also shown for reference. Note that the performance of the localized schemes only varies with time.

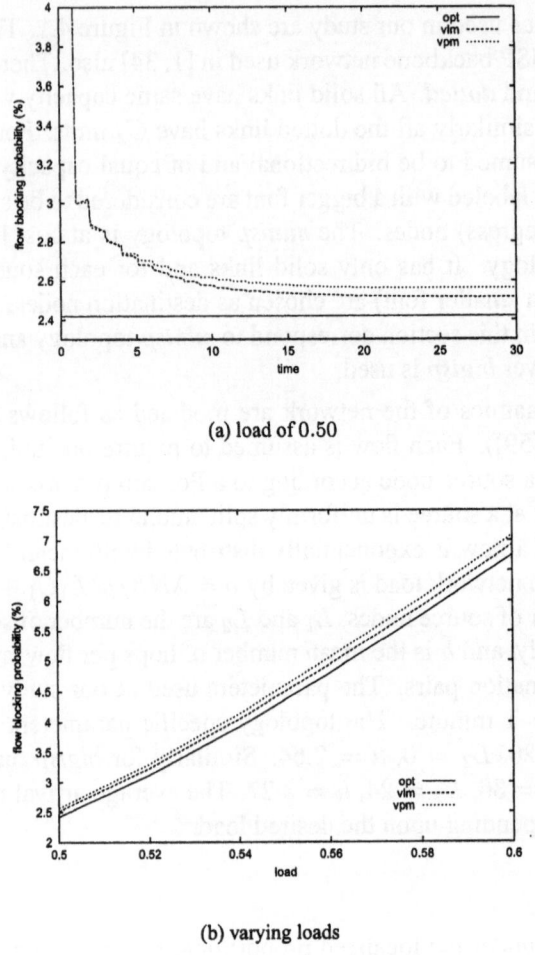

(a) load of 0.50

(b) varying loads

Figure 4.9.   Performance comparison of the optimal and the localized schemes

It can be seen that the overall blocking probability of both the localized routing schemes gradually decreases as the number of iterations increases. Both the schemes eventually converge and each to a different convergence point. Starting with arbitrary initial proportions, the localized schemes approach close to their respective convergence points within 15 iterations. The final convergence points are 2.56%, 2.51% respectively for the schemes *vpm* and *vlm*. The global optimal scheme yields an overall blocking probability of 2.42%. There is very little difference in the performance of finer-grained link-level scheme and the coarser-grained path-level scheme. More importantly, the performance of both the localized schemes is quite close to that of the global optimal scheme.

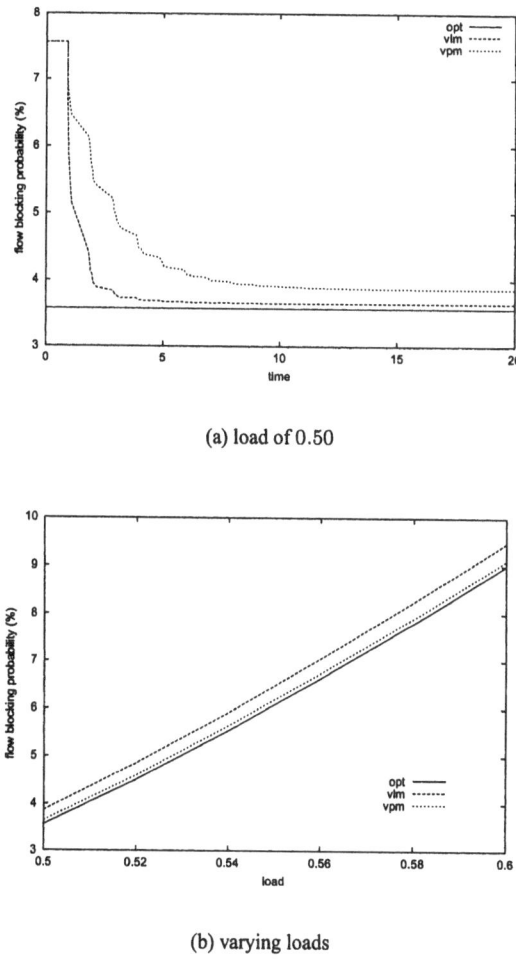

(a) load of 0.50

(b) varying loads

Figure 4.10. Performance under *bigisp* topology

Blocking Performance

The performance of the localized schemes is studied under various load conditions. The load is varied from 0.50 to 0.60. The overall blocking probability at the convergence point of each experiment is plotted as a function of load in Figure 4.9(b). It can be seen that the performance of the *vlm* scheme is slightly better than that of the *vpm* scheme across all loads, and that their performance is quite close to that of the *opt* scheme.

We have also run experiments with the larger *bigisp* topology. These results are shown in Figure 4.10. As before, both the localized virtual capacity based schemes gradually converge to near-optimal proportions even with *bigisp* topology. The blocking at the final convergence points are 3.653% and 3.866% respectively for *vlm* and *vpm* while the optimal scheme yields an overall blocking of 3.57%. Comparatively, these schemes seem to converge relatively sooner in case of *bigisp* than *minisp*. Also, the performance difference between *vlm* and *vpm* is slightly higher in case of *bigisp* than *minisp*. However, the results of *minisp* and *bigisp* are not very different and the observations made with *minisp* are applicable to *bigisp* also. In the next section, we illustrate the effectiveness of localized trunk reservation using *minisp* and show some simulation results on *bigisp* in the later sections.

3.   Alternative Paths and Localized Trunk Reservation

The virtual capacity based local minimization schemes described in the previous section treat all candidate paths equally. Since an admitted flow consumes bandwidth and buffer resources at all the links along a path, clearly path length is also an important factor that must be taken into consideration. There is a fundamental trade-off between minimizing the resource usage by choosing shorter paths and balancing the network load by using lightly loaded longer paths. As a general principle, it is preferable to route a flow along *minhop* (i.e. shortest) paths than paths of longer length (also referred to as *alternative* paths). By preferring minhop paths and discriminating against alternative paths, we not only reduce the overall resource usage but also limit the so-called "knock-on" effect [24, 25], thereby ensuring the stability of the whole system.

The "knock-on" effect refers to the phenomenon where using alternative paths by some sources forces other sources whose minhop paths share links with these alternative paths to also use alternative paths. This cascading effect can cause a drastic reduction of the overall throughput of the network. In order to deal with the "knock-on" effect, trunk reservation [25] is employed where a certain amount of bandwidth on a link is reserved for minhop paths only. With trunk reservation, a flow may be rejected even if sufficient resources are available to accommodate it. A flow along a path longer than its minhop path is admitted only if the available bandwidth even after admitting this flow is greater than the amount of trunk reserved. Trunk reservation provides a simple and yet effective mechanism to control the "knock-on" effect. However, trunk reservation cannot be used directly in localized routing schemes, since it requires global configuration. Furthermore, core routers have to figure out whether a setup request for a flow is sent along its minhop path or not, introducing undesirable burden on them. We propose to address this by having each source router locally discriminate against its own alternative paths *without any explicit global trunk reservation*.

A source node employing a localized scheme can control the amount of alternative routing by adjusting the virtual capacities in its virtual network.

This can be thought of as an implicit localized trunk reservation performed by each source independently. The exact method in which alternative paths are discriminated varies between *vlm* and *vpm*. While *vlm* employs link-level discrimination, *vpm* does path-level discrimination. The details are presented below.

3.1 Localized Link-level Trunk Reservation

Each source router after determining the corresponding virtual private network adjusts the virtual capacities of its links to account for trunk reservation. From the perspective of a source, a link is categorized into two cases: *alternative-only* or *minhop-also*. A link l is said to be *alternative-only* link w.r.t. a source s, if l lies only along alternative paths from the source s to its destinations. Otherwise if a minhop path from source s to any destination passes through link l, then l is categorized as *minhop-also* w.r.t. source s. The links that are used only by alternative paths for routing traffic from this source are targeted for the adjustment. Their capacities are reduced by an amount ψ where, ψ is the trunk reservation parameter, i.e., $vc_{s,l} = (1 - \psi)vc_{s,l}$ if l is *alternative-only* w.r.t. source s. The capacities of other links are left unchanged. The modified *vlm* procedure that incorporates localized trunk reservation is shown in Figure 4.11(a). The virtual capacities for *alternative-only* links are adjusted in lines 4-5. The rest of the procedure remains unchanged.

We now present the rationale behind the way a source under *vlm* categorizes a link as *alternative-only* and applies trunk reservation on it. Consider the three possible cases that a link l can fall into: 1) *minhop-also* w.r.t. to s; 2) *alternative-only* w.r.t. to s and *minhop-also* w.r.t. some other source; 3) *alternative-only* w.r.t. to all sources. We address each of these cases separately as follows

Case 1: No explicit discrimination against alternative paths is required in this case since minimization procedure at source s is expected to account for resource usage while dealing with any sharing of this link l by an alternative path. An alternative path would be assigned an higher capacity on link l only if the minhop path is relatively bad due to other links in its path.

Case 2: By locally reducing the target virtual capacity of link l, source s voluntarily backs off from such links and avoids knock-on effect. This allows minhop paths of other sources to gradually occupy more share on link l. This reduces the resource usage and in turn decreases the overall blocking probability.

Case 3: In this case all the sources reduce the target virtual capacity of link l even though it may not be necessary. This could lead to under-utilization of the link. However, the extent of under-utilization of such links can be

1. PROCEDURE VLM(s)
2.   For each link $l \in L$
3.   Compute virtual capacity $vc_{s,l}^{(n)} = E^{-1}(\nu_{s,l}^{(n)}, b_{s,l}^{(n)})$
4.   For each link $l \in L_s^{alt}$
5.   Decrease virtual capacity $vc_{s,l}^{(n)} = (1 - \psi)vc_{s,l}^{(n)}$
6.   For each path $r \in R_s$
7.   Assign new load $\nu_r^{(n+1)}$ such that
8. $\sum_{r \in R_s} \nu_r^{(n+1)}(1 - b_r)$ is maximum, where
9. $b_r = 1 - \prod_{l \in r}(1 - b_l)$
10. $b_l = E(\nu_{s,l}^{(n+1)}, vc_{s,l}^{(n)})$
11. $\nu_{s,l}^{(n+1)} = \sum_{r \in R_s : l \in r} \nu_r^{(n+1)} \prod_{m \in r - \{l\}}(1 - b_m)$
12. $\sum_{r \in R_\sigma} \nu_r^{(n+1)} = \nu_\sigma, \forall \sigma$
13. END PROCEDURE

(a) *vlm* procedure at source node s

1. PROCEDURE VPM(σ)
2.   For each path $r \in R_\sigma$
3.   Compute virtual capacity $vc_r^{(n)} = E^{-1}(\nu_r^{(n)}, b_r^{(n)})$
4.   For each path $r \in R_\sigma^{alt}$
5.   Decrease virtual capacity $vc_r^{(n)} = (1 - \psi)vc_r^{(n)}$
6.   For each path $r \in R_\sigma$
7.   Find new load $\nu_r^{(n+1)}$ such that
8. $\sum_{r \in R_\sigma} \nu_r^{(n+1)} E(vc_r^{(n)}, \nu_r^{(n+1)})$ is minimum
9. $\sum_{r \in R_\sigma} \nu_r^{(n+1)} = \nu_\sigma$
10. END PROCEDURE

(b) *vpm* procedure for a pair σ

Figure 4.11.   The localized schemes with implicit trunk reservation

limited by setting a low value for trunk reservation parameter ψ since it can be shown that knock-on effect can be avoided even with reasonably small ψ values.

3.2   Localized Path-level Trunk Reservation

The virtual path based scheme limits the extent of alternative routing by applying path-level discrimination against alternative paths. It locally adjusts the target virtual capacity of alternative paths. Given a trunk reservation parameter ψ, the target virtual capacity of alternative paths is reduced by an amount ψ, i.e., $vc_r = (1 - \psi)vc_r$ if r is an alternative path. The virtual capacities of minhop paths are left unchanged. The minimization procedure is then applied

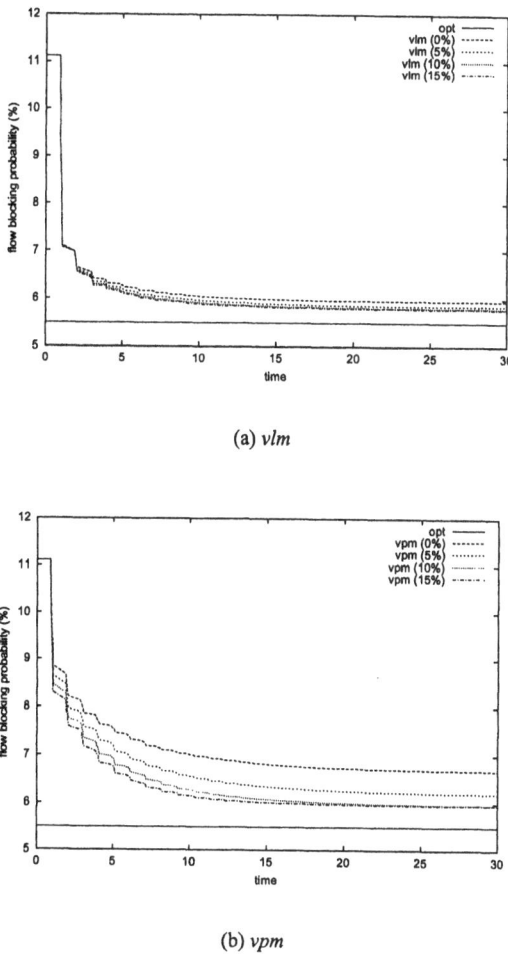

Figure 4.12. The effectiveness of localized trunk reservation (load = 0.60)

locally at the source on the virtual network with these adjusted capacities. The revised *vpm* procedure is shown in Figure 4.11(b). The virtual capacities for alternative paths are adjusted in lines 4-5.

Once again this discrimination against alternative paths helps avoid the knock-on effect. As described in link-level case, when an alternative path shares a link with a minhop path, in path-level case also this implicit trunk reservation by a source helps minhop paths capture more share forcing alternative paths to gradually back off. In other words, the virtual capacity of an alternative path keeps reducing when it shares a bottleneck link with a minhop path. On the other hand when an alternative path doesn't share any bottleneck links with any minhop path, the amount of reduction in its target virtual capacity will be limited by the trunk reservation parameter ψ thus limiting the amount of under-utilization of resources.

3.3　Effectiveness of Localized Trunk Reservation

We now study the effectiveness of localized trunk reservation method and the impact of the parameter ψ on the performance. The setup described in Section 4.2.4.0 is used in this study also. However, apart from the minhop paths, paths of length minhop+1 are also chosen as the candidate paths for each source-destination pair.

The Figure 4.12 shows the convergence process where overall blocking probability is plotted as a function of time for a load of 0.60. The performance of local schemes is shown for four different values of the trunk reservation parameter ψ: 0%,5%,10%,15%. The blocking probability of global optimal scheme is shown for reference. It is quite evident that the performance of localized schemes with ($> 0\%$) trunk reservation is better than without (0%) it. However, as the ψ value is increased the performance gain is reduced. There is almost no difference in performance between ψ values of 10% and 15%. We also ran for several cases with loads ranging from 0.50 to 0.70. In each case the localized schemes were run with ψ values of 0% and 10%. The Figure 4.13 shows the overall blocking probability as a function of load. Once again, across all loads, the performance of schemes with localized trunk reservation is better than without it.

These results show that localized trunk reservation is quite effective. However, the impact of localized trunk reservation on *vlm* is much less significant than on *vpm*. This is expected since *vlm* even without any trunk reservation, using finer-grained link-level information, accounts for sharing of links between minhop and alternative paths from a source to all its destinations. The role of localized trunk reservation in *vlm* is limited to avoiding overloading of minhop paths of a source by traffic on alternative paths of another source. On the other hand, the localized trunk reservation plays a much more critical role in *vpm*. The minhop paths to a destination have to be guarded from alternative paths to the same destination besides from alternative paths to other destinations. This is due to availability of only coarser-grain information and thus lack of knowledge about sharing of links between different paths. However with localized trunk reservation, the *vpm* scheme tides over this shortcoming and performs comparably to the *vlm* scheme. Hereafter, we focus only on the proportional routing schemes such as *vpm* that are based on path-level information which is easier to collect with less overhead than link-level information.

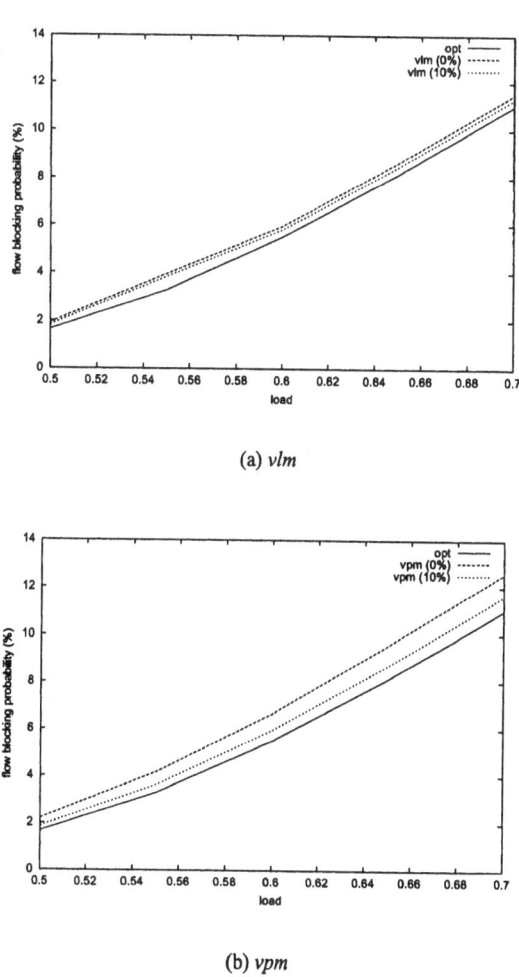

(a) *vlm*

(b) *vpm*

Figure 4.13. The effectiveness of localized trunk reservation

Chapter 5

LOCALIZED PROPORTIONAL ROUTING: PRACTICAL SCHEMES

The localized schemes described in the previous chapter have been shown to approach the performance of the optimal scheme using only local information. Furthermore, even with coarser-grain path-level blocking information the *vpm* scheme performs as well as the *vlm* scheme that uses finer-grain link-level blocking information. It is easier to collect path-level statistics and simpler to implement path-based schemes. Hence we focus on path-based localized schemes and further investigate the issues involved in implementing them.

1. Heuristic Equalization Strategies

The *vpm* scheme first computes the virtual capacity of each candidate path and then performs local minimization. Though the complexity of this minimization procedure is much less than that of optimal scheme, it could still be significant. We are interested in simple schemes that are easy to implement. A simple alternative to *minimization* procedure is *equalization* of either blocking probabilities or blocking rates of candidate paths.

1.1 Equalization of Blocking Probabilities

The objective of the equalization of blocking probabilities (*ebp*) strategy is to find a set of proportions $\{\tilde{\alpha}_1, \tilde{\alpha}_2, \ldots, \tilde{\alpha}_k\}$ such that flow blocking probabilities of all the paths are equalized, i.e., $\tilde{b}_1 = \tilde{b}_2 = \cdots = \tilde{b}_k$, where $\tilde{b}_i$ is the flow blocking probability of path r_i, and is given by $E(\tilde{\alpha}_i \nu, c_i)$.

1.2 Equalization of Blocking Rates

The objective of the equalization of blocking rates (*ebr*) strategy is to find a set of proportions $\{\hat{\alpha}_1, \hat{\alpha}_2, \ldots, \hat{\alpha}_k\}$ such that *flow blocking rates* of all the

1.	PROCEDURE VCEBP(σ)
2.	Set mean blocking rate of minhop paths, $\bar{\beta}^{(n)} = \sum_{r \in R_\sigma^{min}} \alpha_r^{(n)} b_r^{(n)}$
3.	Set minimum of minhop path's blocking probability, $b^* = \min_{r \in R_\sigma^{min}} b_r^{(n)}$
4.	For each path $r \in R_\sigma^{min}$
5.	Compute virtual capacity $vc_r^{(n)} = E^{-1}(\nu_r^{(n)}, b_r^{(n)})$
6.	For each path $r \in R_\sigma^{min}$
7.	Compute target load ν_r^* such that $\bar{\beta}^{(n)} = E(\nu_r^*, vc_r^{(n)})$
8.	For each alternative path $r \in R_\sigma^{alt}$
9.	Compute target load ν_r^* such that $(1 - \psi)b^* = E(\nu_r^*, vc_r^{(n)})$
10.	For each path $r \in R_\sigma$
11.	Compute new proportion $\alpha_r^{(n+1)} = \dfrac{\nu_r^*}{\sum_{r \in R_\sigma} \nu_r^*}$
12.	END PROCEDURE

Figure 5.1. The *vcebp* procedure for a source-destination pair σ

paths are equalized, i.e., $\hat{\alpha}_1 \hat{b}_1 = \hat{\alpha}_2 \hat{b}_2 = \cdots = \hat{\alpha}_k \hat{b}_k$, where $\hat{b}_i$ is the flow blocking probability of path r_i, and is given by $E(\hat{\alpha}_i \nu, c_i)$.

The proportions corresponding to the above equalization strategies can be computed using an iterative procedure similar to the *vpm* procedure. The key difference is in the way new proportions for paths are computed based on their current virtual capacities in each iteration. While *vpm* tries to minimize the aggregate blocking probability for flows between a source-destination pair, equalization strategies attempt to equalize the probability or rate of blocking experienced by flows across different candidate paths between the pair.

The virtual capacity based equalization of blocking probabilities (*vcebp*) procedure is shown in Figure 5.1. At any given iteration $n \geq 0$, let $\nu_r^{(n)}$ be the amount of the load currently routed along a path $r \in R_\sigma$, and let $b_r^{(n)}$ be its observed blocking probability on the path. Then the virtual capacity of path r is given by $vc_r = E^{-1}(\nu_r^{(n)}, b_r^{(n)})$ (line 5). For each minhop path, the mean blocking probability of all the minhop paths, $\bar{\beta}^{(n)}$, is used to compute a new target load (lines 6-7). Similarly, for each alternative path, a new target load is computed using the target blocking probability $(1 - \psi)b^*$ (lines 8-9). Here b^* is the minimum flow blocking probability of all the minhop paths (line 3) and ψ is a configurable parameter to limit the knock-on effect. The basic idea behind this alternative routing method is to ensure that an alternative path is used to route flows between the source-destination pair only if it has a "better quality" (measured in flow blocking probability) than any of the minhop paths. Given these new target loads for all the paths, the new proportion of flows, $\alpha_r^{(n+1)}$, for each path r is obtained in lines 10-11, resulting in a new load $\nu_r^{(n+1)} = \alpha_r^{(n+1)} \nu_\sigma$ on path r. Similar procedure can be used to equalize the

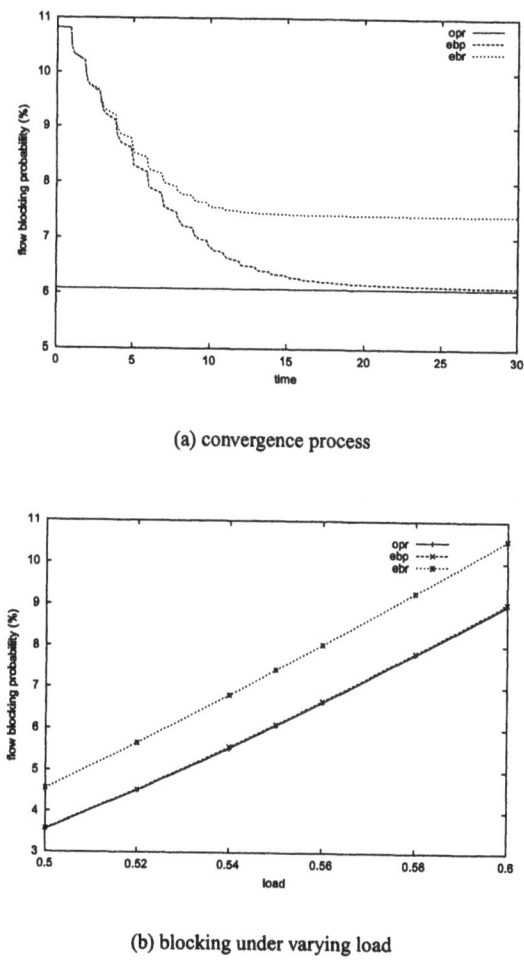

(a) convergence process

(b) blocking under varying load

Figure 5.2.   Performance of *ebp* and *ebr*

blocking rates also by computing mean blocking rate in line 2 and using that
as the target rate in line 7 for computing the target loads.

Figure 5.2 compares the performance of *ebp* and *ebr* with *opr*. We consider
the ideal case where the observed blocking probabilities on paths are precisely
computed numerically. These experiments are conducted on the *bigisp* topology shown in Figure 4.1(a) in the previous chapter. In these experiments only
the minhop paths are made candidates and the load is set to 0.55. Figure 5.2(a)
shows the convergence of the proposed heuristic schemes. We can see that *ebp*
scheme converges to almost optimal proportions. The *ebr* scheme also converges but its blocking performance is worse than *ebp*. Figure 5.2(b) shows the

blocking performance of these schemes under various offered loads. Across all loads, there is no discernible difference in the performance between *opr* and *ebp*. On the other hand, *ebr* performs consistently worse than *ebp*. The reason is that *ebr* scheme assigns more load to a higher blocking path than *ebp* scheme in an attempt to equalize the blocking rates of candidate paths. This effect gets amplified when there is lot of sharing of links between paths. However, *ebr* is a reasonable strategy that effectively assigns loads to paths in proportions that are inversely proportional to their blocking probabilities. In the following we proceed to explore practical implementations of both these strategies.

The localized routing schemes presented so far are based on theoretical virtual capacity model. We have shown that they yield near-optimal performance using only local information. However, there are two difficulties involved in implementing the virtual capacity model. First, computation of virtual capacity and target load using Erlang's Loss Formula can be quite cumbersome. Second, and perhaps more importantly, the accuracy in using Erlang's Loss Formula to compute virtual capacity and new load relies critically on steady-state observation of flow blocking probability. Hence small statistic variations may lead to erroneous flow proportioning, causing undesirable load fluctuations. In order to circumvent these difficulties, we are interested in simple yet robust implementation of these schemes. In the following we discuss two such schemes that approximate *ebr* and *ebp* respectively. We first present the *psr* scheme that we designed initially and show that it is viable alternative to popular global best-path routing scheme *wsp*. We then describe an approximation of *ebp* that is found to perform even better than *psr*.

2.   Proportional Sticky Routing

The proportional sticky routing (*psr*) scheme attempts to equalize blocking rates of candidate paths. It is called by that name because it essentially does proportional routing while obtaining proportions through a form of sticky routing. The *psr* scheme can be viewed to operate in two stages: 1) proportional flow routing, and 2) computation of flow proportions. The proportional flow routing stage proceeds in *cycles* of variable length. During each cycle incoming flows are routed along paths selected from a set of eligible paths. A path is selected with a frequency determined by a prescribed proportion. A number of cycles form an *observation period*, at the end of which a new flow proportion for each path is computed based on its observed blocking probability. This is the computation of flow proportion stage. The flow proportions for min-hop paths of a source-destination pair are determined using the *ebr* strategy, whereas flow proportions for alternative paths are determined using a target blocking probability. In the following we will describe these two stages in more detail.

1.	PROCEDURE PSR-ROUTE()
2.	Select an eligible path $r = wrrps(R^{elg})$
3.	Increment flow counter, $n_r = n_r + 1$
4.	If failed to setup connection along r
5.	Decrement failure counter, $f_r = f_r - 1$
6.	If failures reached limit, $f_r == 0$
7.	Remove r from eligible set, $R^{elg} = R^{elg} - r$
8.	If eligible set is empty, $R^{elg} == \emptyset$
9.	Reset eligible set, $R^{elg} = R$
10.	For each path $r \in R$
11.	Reset failure counter, $f_r = \gamma_r$
12.	END PROCEDURE

(a) proportional routing

1.	PROCEDURE PSR-PROPO-COMPU()
2.	For each path $r \in R$
3.	Compute blocking probability, $b_r = \frac{\eta\gamma_r}{n_r}$
4.	Assign a proportion, $\alpha_r = \frac{n_r}{\sum_{\hat{r} \in R} n_{\hat{r}}}$
5.	Set target blocking probability, $b^* = \min_{r \in R^{min}} b_r$
6.	For each alternative path $r' \in R^{alt}$
7.	If blocking probability *high*, $b_{r'} >= b^*$
8.	Decrement failure limit, $\gamma_{r'} = \gamma_{r'} - 1$
9.	If blocking probability *low*, $b_{r'} < \psi b^*$
10.	Increment failure limit, $\gamma_{r'} = \gamma_{r'} + 1$
11.	END PROCEDURE

(b) computation of proportions

Figure 5.3. The *psr* procedure

2.1 Proportional flow routing

Given an arbitrary source-destination pair, let R be the set of (*explicit-routed*) paths between the source-destination pair, where $R = R^{min} \cup R^{alt}$. We associate with each path $r \in R$, a *maximum permissible flow blocking* parameter γ_r and a corresponding *flow blocking counter* f_r. For each minhop path $r \in R^{min}$, $\gamma_r = \hat{\gamma}$, where $\hat{\gamma}$ is a configurable system parameter. For each alternative path $r' \in R^{alt}$, the value of $\gamma_{r'}$ is dynamically adjusted between 1 and $\hat{\gamma}$, as will be explained later. As shown in Figure 5.3(a), at the beginning of each cycle, f_r is set to γ_r. Every time a flow routed along path r is blocked, f_r is decremented. When f_r reaches zero, path r is considered *ineligible*. At any time only the set of *eligible* paths, denoted by R^{elg}, is used to route flows. A path from current eligible path set R^{elg} is selected using a weighted-round-robin-like path selector (*wrrps*). The *wrrps* procedure is described below. Once R^{elg} becomes empty, the current cycle is ended and a new cycle is started with $R^{elg} = R$ and $f_r = \gamma_r$.

Weighted Round Robin Procedure for Path Selection

Given a set R^{elg} of eligible paths and their associated proportions $\{\alpha_r, r \in R^{elg}\}$, *wrrps* picks a path $r \in R^{elg}$ based on its weight, $w_r = \frac{\alpha_r}{\sum_{s \in R^{elg}} \alpha_s}$. Instead of using a probabilistic method such as picking a path r with probability w_r, we opt to employ a deterministic algorithm to ensure that flow proportions are preserved within as small a time window as possible. This is implemented by using a deterministic sequence of paths which has the property that the paths are distributed periodically with a frequency which closely approximates the prescribed flow proportions. This sequence is generated by *wrrps* on the fly: for an incoming flow, *wrrps* generates the next path in the sequence and routes the flow along the path.

The *wrrps* procedure is shown in Figure 5.4. It keeps track of the number of times each path was selected (n_{r_i}) and the run length (l) of the most recently selected path. It maintains an ordered list of paths and the first path in the list is selected as long as it satisfies both the following constraints: 1) its weight is more than its run length times the weight of the rest of paths ($l \mathcal{W}_1 < w_{r_0}$); 2) ratio of the number of times it was selected and the number of times all others were selected is less than or equal to the ratio of its weight and weight of the rest of paths ($\mathcal{W}_1 n_{r_0} \le w_{r_0} \mathcal{N}_1$). Otherwise this path is pushed down the order and the run length is reset to 0. Then it returns the first path in the list. A sample *wrrps* generated sequence where the current eligible set R^{elg} has four paths r_1, r_2, r_3 and r_4 with weights $1/2, 1/4, 1/8$, and $1/8$ respectively is: $r_1\ r_2\ r_1\ r_3\ r_1\ r_2\ r_1\ r_4\ r_1\ r_2\ r_1\ \mathbf{r_3}$. This sequence has the property that in every window of size 2 there is an r_1 and an r_2 in every window of size 4. Similarly, one r_3 and one r_4 in all windows of size 8. Assuming that up to the last r_1 are the paths chosen so far, the next path selected on the fly by the *wrr* path selector would be r_3. Note also that every time the eligible path set R^{elg} changes, a new sequence is generated, and flows arriving thereafter are thus routed according to this new sequence.

2.2 Computation of flow proportions

Flow proportions $\{\alpha_r, r \in R\}$ are recomputed at the end of each observation period (see Figure 5.3(b)). An observation period consists of η cycles, where η is a configurable system parameter used to control the robustness and stability of flow statistics measurement. During each observation period, we keep track of the number of flows routed along each path $r \in R$ using a counter n_r. At the beginning of an observation period, n_r is set to 0. Every time path r

k	:	total number of paths in set R^{elg}.
r_i	:	path associated with index i.
w_{r_i}	:	weight associated with path r_i.
n_{r_i}	:	number of times path r_i was selected.
l	:	run length of the most recently selected path.
$\mathcal{W}_i$	:	$w_{r_i} + w_{r_{i+1}} + \cdots + w_{r_k}$.
$\mathcal{N}_i$	:	$n_{r_i} + n_{r_{i+1}} + \cdots + n_{r_k}$.

(a) notation

```
1.   PROCEDURE wrrps()
2.     For i = 1, 2, . . . , k
3.       If lW_{i+1} < w_{r_i} and W_{i+1}n_{r_i} ≤ w_{r_i}N_{i+1}
4.         break
5.       Set W_{i+1} = W_{i+1} + w_{r_i} - w_{r_{i+1}}
6.       Set N_{i+1} = N_{i+1} + n_{r_i} - n_{r_{i+1}}
7.       Swap r_i and r_{i+1}
8.     Set l = 0
9.     Set n_{r_0} = n_{r_0} + 1; N_{r_0} = N_{r_0} + 1;
10.    Set l = l + 1
11.    Return r_0
12.  END PROCEDURE
```

(b) path selection

Figure 5.4.   The *wrrps* procedure

is used to route a flow, n_r is incremented. Since an observation period consists of η cycles, and in every cycle, each path r has exactly γ_r flows blocked, the observed flow blocking probability on path r is $b_r = \frac{\eta\gamma_r}{n_r}$. For each minhop path $r \in R^{min}$, its new proportion α_r is recomputed at the end of an observation period and is given by $\alpha_r = n_r/n_{total}$, where $n_{total} = \sum_{r \in R} n_r$ is the total number of flows routed during an observation period. Recall that for a minhop path $r \in R^{min}$, $\gamma_r = \hat{\gamma}$. Hence $\alpha_r b_r = \frac{n_r}{n_{total}} \frac{\eta\gamma_r}{n_r} = \frac{n_r}{n_{total}} \frac{\eta\hat{\gamma}}{n_r} = \frac{\eta\hat{\gamma}}{n_{total}}$. This shows that the above method of assigning flow proportions for the minhop paths equalizes their flow blocking rates.

We use the minimum blocking probability among the minhop paths, $b^* = \min_{r \in R^{min}} b_r$, as the reference to control flow proportions for the alternative paths. This is done implicitly by dynamically adjusting the maximum permissible flow blocking parameter $\gamma_{r'}$ for each alternative path $r' \in R^{alt}$. At the end of an observation period, let $b_{r'} = \frac{\eta\gamma_{r'}}{n_{r'}}$ be the observed flow blocking probability for an alternative path r'. If $b_{r'} > b^*$, $\gamma_{r'} := \max\{\gamma_{r'} - 1, 1\}$. If $b_{r'} < \psi b^*$, $\gamma_{r'} := \min\{\gamma_{r'} + 1, \hat{\gamma}\}$. If $\psi b^* \leq b_{r'} \leq b^*$, $\gamma_{r'}$ is not changed.

By having $\gamma_{r'} \geq 1$, we ensure that some flows are occasionally routed along alternative path r' to probe its "quality", whereas by keeping $\gamma_{r'}$ always below $\hat{\gamma}$, we guarantee that minhop paths are always preferred to alternative paths in routing flows. The new proportion for each alternative path r' is again given by $\alpha_{r'} = n_{r'}/n_{total}$. Note that since $\gamma_{r'}$ is adjusted for the next observation period, the *actual* number of flows routed along alternative path r' will be also adjusted accordingly.

2.3   Performance Evaluation and Analysis

This section evaluates the performance of the proposed localized proportional routing scheme *psr* and compares it with the global best-path routing scheme *widest shortest path* (*wsp*). We start with the description of the simulation environment and then compare the performance of *psr* and *wsp* in terms of the overall blocking probability and routing overhead.

Simulation Environment

Figure 5.5 shows the two topologies, *bigisp* and *rand*, used in our study. The *bigisp* topology is same as the one in Figure 4.1(a). However, the simulation setting here is somewhat different from the one described in Section 4.2.4.0. Hence we detail the simulation environment again though there is quite a bit of overlap in the settings. The *rand* topology is a random graph generated by GT-ITM [68] and used in [19]. For simplicity, all the links are assumed to be bidirectional and of equal capacity in each direction. The *rand* topology has three types of links: *thin*, *thick* and *dotted* while *bigisp* topology has only *thin* links. All *thin* links have same capacity with C_1 units of bandwidth and similarly all the *thick* links have C_2 units. The *dotted* links are the access links and for the purpose of our study their capacity is assumed to be infinite. Flows arriving into the network are assumed to require one unit of bandwidth. Hence a link with capacity C can accommodate at most C flows simultaneously.

The flow dynamics of the network is modeled as follows (similar to the model used in [59]). Flows arrive at a source node according to a Poisson process with rate λ. The destination node of a flow is chosen randomly from the set of all nodes except the source node. In case of *bigisp* all nodes are considered to be capable of being source or destination nodes for flows. But in case of *rand* topology, only the nodes attached to the *dotted* access links are assumed to be end points of flows. The holding time of a flow is exponentially distributed with mean $1/\mu$. Following [59], the offered network load on *bigisp* is given by $\rho = \lambda N \bar{h}/\mu L_1 C_1$, where N is the number of source nodes, L_1 the number of links, and $\bar{h}$ is the mean number of hops per flow, averaged across all source-destination pairs. Similarly the offered load on *rand* is given by $\rho = \lambda N \bar{h}/\mu (L_1 C_1 + L_2 C_2)$, where L_1 and L_2 are the number of *thin* and *thick* links respectively. The parameters used in simulation are $C_1 = 20$,

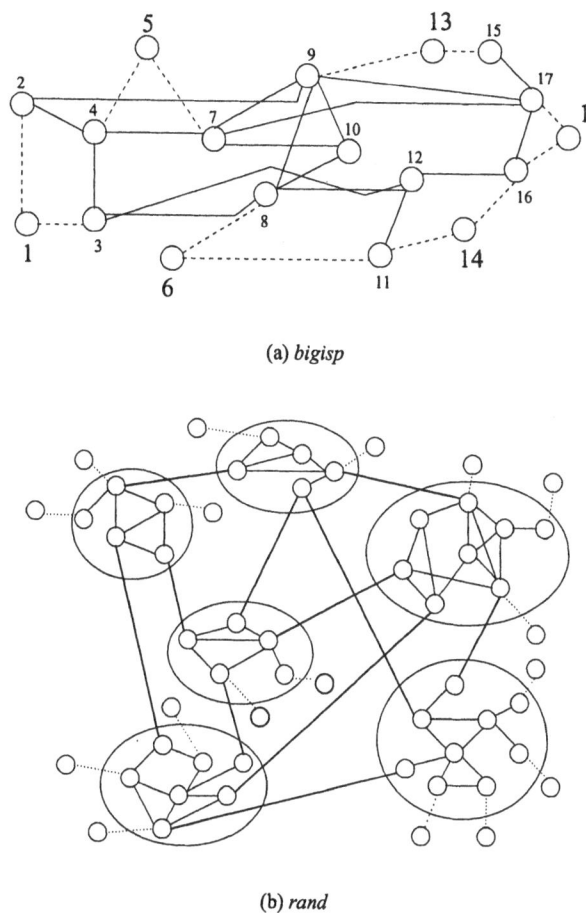

(a) *bigisp*

(b) *rand*

Figure 5.5.   Topologies used in performance evaluation

$C_2 = 40$, $1/\mu = 60$ sec. The topology specific parameters are $N = 18$, $L_1 = 60$, $\bar{h} = 2.36$ for *bigisp* and $N = 56$, $L_1 = 100$, $L_2 = 22$, $\bar{h} = 4.38$ for *rand*. The average arrival rate at a source node λ is set depending upon the desired load.

Blocking Probability

The performance of *wsp* and *psr* is compared by measuring the blocking probability under various settings. We first present the impact of update interval on the performance of *wsp* and show how the blocking probability increases rapidly as update interval is increased. We then compare the performance under various loads and conclude that the blocking probability of *psr* is compa-

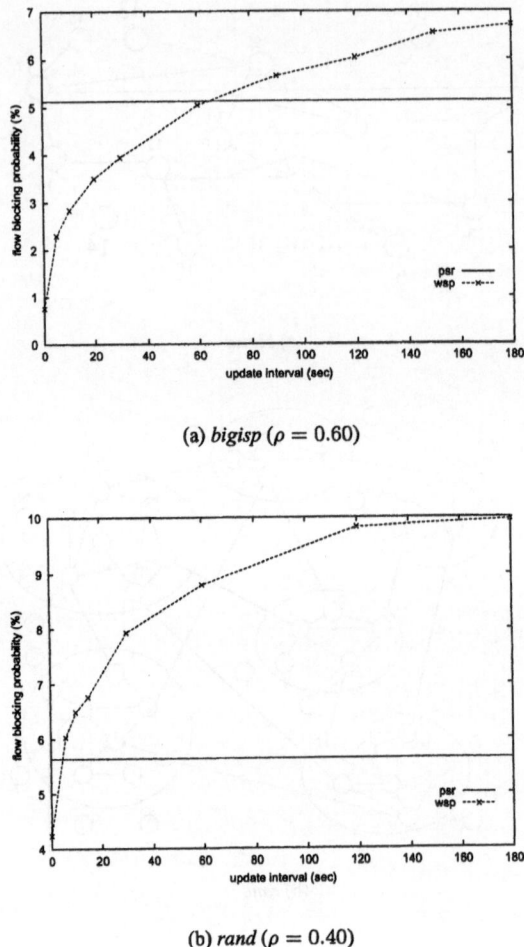

(a) *bigisp* ($\rho = 0.60$)

(b) *rand* ($\rho = 0.40$)

Figure 5.6.   Impact of update interval

rable to that of *wsp* even at small update intervals. Finally we measure their performance under non-uniform load conditions and demonstrate that *psr* is better at alleviating the effect of "hot spots".

Varying update interval

Figure 5.6 compares the performance of *wsp* and *psr* for both *bigisp* and *rand* topologies. The values for configurable parameters in *psr* were set to $\eta = 3$, $\hat{\gamma} = 5$, and $\psi = 0.8$. The performance is measured in terms of the over-all flow blocking probability, which is defined as the ratio of the total number of blocks to the total number of flow arrivals. The overall blocking probability is plotted as a function of the update interval. The offered load was set to 0.60

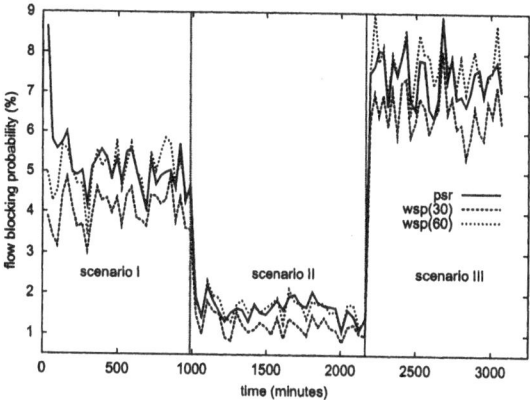

Figure 5.7. Performance under varying uniform load

in case of *bigisp* and 0.40 in case of *rand*. Note that update interval is used only in *wsp* in conjunction with a threshold based triggering policy, to enforce a minimum spacing between updates. The threshold based policies trigger an update whenever the percent of change in bandwidth is greater than a constant threshold value. In our simulations, this threshold is set to 50%[1]. From the figures, we see that as the update interval of *wsp* increases, the blocking probability of *wsp* rapidly approaches that of *psr* and performs worse for larger update intervals. In the case of *bigisp* topology, *psr* performs better than *wsp* when the update interval goes beyond 65 sec. For *rand* topology, this crossover happens at a much smaller update interval of less than 10 sec. This shows that *psr* using only local information performs better than global information exchange based *wsp* even when the update interval is reasonably low.

Varying offered load

We now illustrate the adaptivity of *psr* by varying the offered load on *bigisp*. We initially offer a load of 0.60 as was done in the earlier simulation and then this overall load is decreased to 0.50 and again increased to 0.65. We plot the blocking probability under *psr* and *wsp* as a function of time in Figure 5.7. The performance of *wsp* is shown for two update intervals: 30 sec and 60 sec. Starting with arbitrary initial proportions, *psr* quickly converges and performs as well as *wsp*(60). When the load is decreased, *psr* adapts to the change and maintains its relative performance. Finally, when the load is increased to 0.65, once again it reacts promptly and performs slightly better than *wsp*(60).

---

[1]Note that blocking performance of threshold based trigger with hold-down timer T would be no better than simple timer based trigger with update interval of T. The key difference is in the amount of update message overhead.

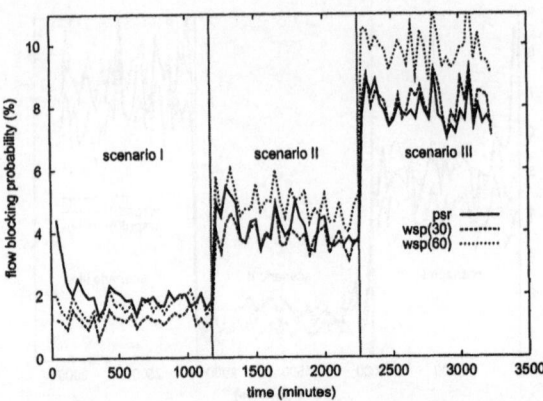

Figure 5.8.   Performance under varying non-uniform load

Varying non-uniform traffic

It is likely that a source node receives a larger number of flows to a few specific destinations [9], i.e, a few destinations are "hot". Ideally a source would like to have more up-to-date view of the QoS state of the links along the paths to these "hot" destinations. In the case of *wsp*, this requires more frequent QoS state updates, resulting in increased overhead. But in the case of *psr*, because of its adaptivity and statistics collection mechanism, a source does have more accurate information about the frequently used routes and thus alleviates the effect of "hot spots". We illustrate this by introducing increased levels of traffic between certain pairs of network nodes ("hot pairs") in *bigisp*, as was done in [1]. Apart from the normal load that is distributed between all source-destination pairs, an additional load (hot load) is distributed among all the hot pair nodes. The hot pairs chosen are $(2, 16)$, $(3, 17)$, and $(9, 11)$.

We consider three scenarios. In scenario I, a load of 0.50 is offered uniformly among all the nodes as was done in earlier simulations. In scenario II, an additional load of 0.05 is offered between hot pairs only and in scenario III this additional load is further increased to 0.10. Figure 5.8 shows the blocking performance of the two schemes under different scenarios as a function of time. Under scenario I, starting with arbitrary initial proportions, *psr* quickly converges to a stable state where its blocking probability is similar to that of *wsp*(60). But in scenario II with additional load between hot pairs, *psr* approaches the performance of *wsp*(30) and even better in scenario III where the load between hot pairs is higher. These results illustrate the degradation in performance of *wsp* and improvement in relative performance of *psr* under non-uniform load conditions.

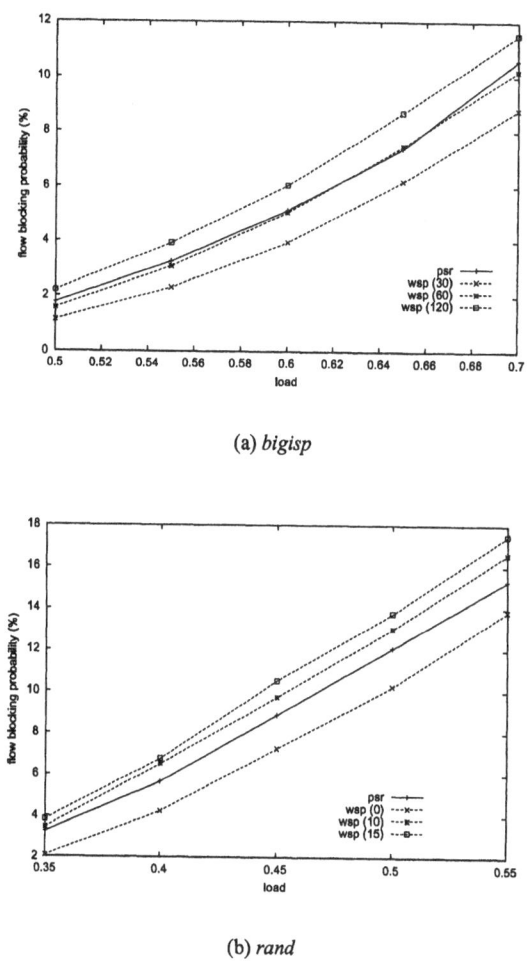

(a) *bigisp*

(b) *rand*

Figure 5.9. Performance under various loads

Various Load Conditions

We have further investigated the impact of traffic pattern on the relative performance of these schemes by offering various loads. First, we consider the setting where load is offered uniformly between all the nodes. Figure 5.9 shows the blocking performance of these two schemes as a function of the offered network load. As before, the performance is measured in terms of the overall flow blocking probability. The network load is varied from 0.50 to 0.70 in case *bigisp* and 0.35 to 0.55 in case of *rand*. The performance of *wsp* is plotted for three update intervals of 30, 60 and 120 for the *bigisp* case and similarly for 0, 10, and 15 in case of *rand*. It is clear that the blocking performance of *psr*

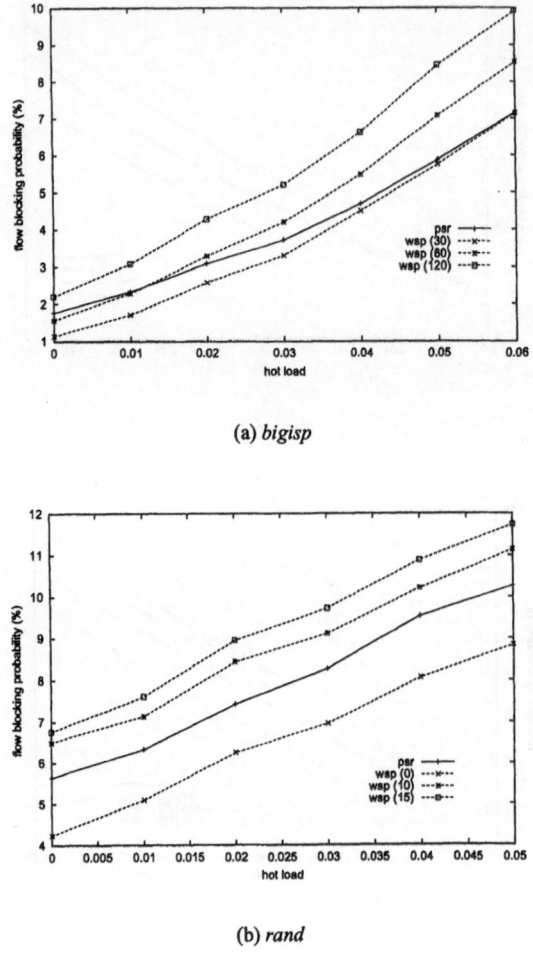

(a) *bigisp*

(b) *rand*

Figure 5.10. Performance under non-uniform load conditions

is quite similar to that of *wsp* with update interval of 60 sec in case of *bigisp* topology. For the case of *rand* the *psr*'s performance is better than *wsp* with update interval of 10 sec. The performance of *wsp* is poor, particularly in case of *rand* where the paths are longer and the number of *minhop* paths are fewer. In such a case, *wsp* scheme due to its preference for shortest paths, seem to select *minhop* paths even when they are congested. On the other hand, *psr* uses alternate paths judiciously and yields much lower blocking probability. These results indicate that even at smaller update intervals *wsp* fares no better than *psr*.

We then fixed the normal load that is distributed between all source destination pairs, and varied an additional load (hot load) is distributed among all the hot pair nodes. The hot pairs chosen for *bigisp* topology are $(2, 16)$, $(3, 17)$, and $(6, 15)$. Similarly in case of *rand* topology, the thick routers are considered to be the web servers and hence more traffic is assumed to flow from these two nodes to all other nodes. The normal load is fixed at 0.50 for *bigisp* and 0.40 for *rand* and the extra load is varied. Figure 5.10 shows the overall flow blocking probability as a function of the extra load.

First consider the case of *bigisp* topology. When the additional load is less than 0.01, the blocking performance of *psr* is comparable to *wsp* with update interval of 60 sec. But as the traffic between hot pairs increases, *psr* progressively does better in comparison to *wsp*. Particularly when the hot load is 0.06, the performance of *psr* is even better than *wsp* with an update interval of 30 sec. In case of *rand* topology the performance of *psr* is much better than *wsp* with update interval 10 sec irrespective of the amount of hot load. This not only shows the limitation of global QoS routing schemes such as *wsp* but also illustrates the advantage of self-adaptivity in localized QoS routing schemes such as *psr*.

2.4   Heterogeneous Traffic

The discussion so far is focused on the case where the traffic is homogeneous, i.e., all flows request for one unit of bandwidth and their holding times are derived from the same exponential distribution with a fixed mean value. Here we study the applicability of *psr* in routing heterogeneous traffic where flows could request for varying bandwidths with their holding times derived from different distributions. We demonstrate that *psr* is insensitive to the duration of individual flows and hence we do not need to differentiate flows based on their holding times. We also show that when the link capacities are considerably larger than the average bandwidth request of flows, it may not be necessary to treat them differently and hence *psr* can be used *as is* to route heterogeneous traffic.

Consider the case of traffic with k types of flows, each flow of type i having a mean holding time $1/\mu_i$ and requesting bandwidth B_i. Let ρ_i be the offered load on the network due to flows of type i, where the total offered load, $\rho = \sum_{i=1}^{k} \rho_i$. The fraction of total traffic that is of type i, $\phi_i = \rho_i/\rho$. The arrival rate of type i flows at a source node, λ_i is given by $\lambda_i = \rho_i \mu_i LC / N\bar{h}B_i$, which is an extension of the formula presented in Section 5.2.3.0. To account for the heterogeneity of traffic, bandwidth blocking ratio [34] is used as the performance metric for comparing different routing schemes. The bandwidth blocking ratio is defined as the ratio of the bandwidth usage corresponding to blocked flows and the total bandwidth usage of all the offered traffic. Suppose b_i is the observed blocking probability for flows of type i, then the bandwidth blocking ratio is given by $\frac{\sum_{i=1}^{k} \frac{b_i \lambda_i B_i}{\mu_i}}{\sum_{i=1}^{k} \frac{\lambda_i B_i}{\mu_i}}$. In the following, we compare the performance of *psr* and *wsp*, measured in terms of bandwidth blocking ratio, under different traffic conditions, varying the fractions ϕ_i to control the traffic mix.

Mixed holding times

We now examine the case of traffic with 2 types of flows that request for the same amount of bandwidth, i.e, $B_1 = B_2 = 1$, but with different holding times. We consider three scenarios. In the first scenario, both types of flows have their holding times derived from exponential distribution but their means are different: 60 and 120 sec. In the second scenario, both types have the same mean holding time of 60 sec but their distributions are different: exponential and pareto. In the third scenario, holding times of both types of flows follow pareto distribution but their means are different: 60 and 120 sec. In all these scenarios, a load of 0.40 is offered between the border nodes in *isp*. Figure 5.11 shows the performance of *psr* and *wsp* under different scenarios.

Consider the first scenario where type 1 flows are *short* ($\frac{1}{\mu_1}$ = 60 sec) and type 2 flows are *long* ($\frac{1}{\mu_1}$ = 120 sec), but both are exponentially distributed. Figure 5.11(a) shows the bandwidth blocking ratio plotted as a function of the fraction ϕ_1 corresponding to short flows. It is quite evident that the performance of *wsp* degrades as the proportion of *short* flows increases while that of *psr* stays almost constant. The behavior of *wsp* is as expected since the shorter flows cause more fluctuation in the network QoS state and the information at a source node becomes more inaccurate as the QoS state update interval gets larger relative to flow dynamics. On the contrary, *psr* is insensitive to the duration of flows.

In the second scenario, a fraction of flows have their holding times derived from a pareto distribution while the rest have their holding times derived from an exponential distribution. The mean holding time of both the types is the same, 60 sec. The pareto distribution is long tailed with its tail controlled by a *shape* parameter. We have experimented with different shape values in the range 2.1 to 2.5 and found that results are similar. The results reported here correspond to a shape value of 2.2. In Figure 5.11(b), bandwidth blocking ratio is plotted as a function of the fraction of pareto type flows. As the fraction of pareto flows increases, the blocking under *wsp*(30) increases while it stays almost same under *wsp*(15). The number of short (much less than mean holding time) flows are more under the pareto distribution than the exponential distribution because of the long tail of pareto. Consequently, update interval has to be small to capture the fluctuations due to such short flows. That is why the performance of *wsp*(30) degrades while *wsp*(15) is not affected. The relative performance of these schemes in the third scenario is similar to the first sce-

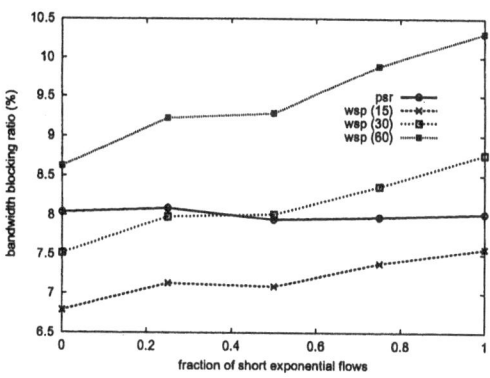

(a) long and short exponential

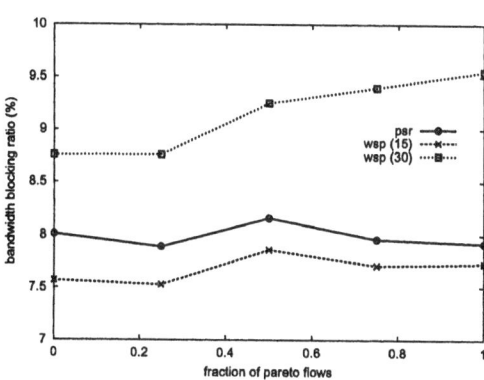

(b) exponential and pareto

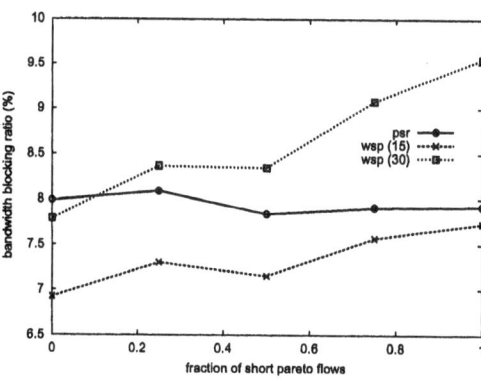

(c) long and short pareto

Figure 5.11. Traffic with mixed holding times

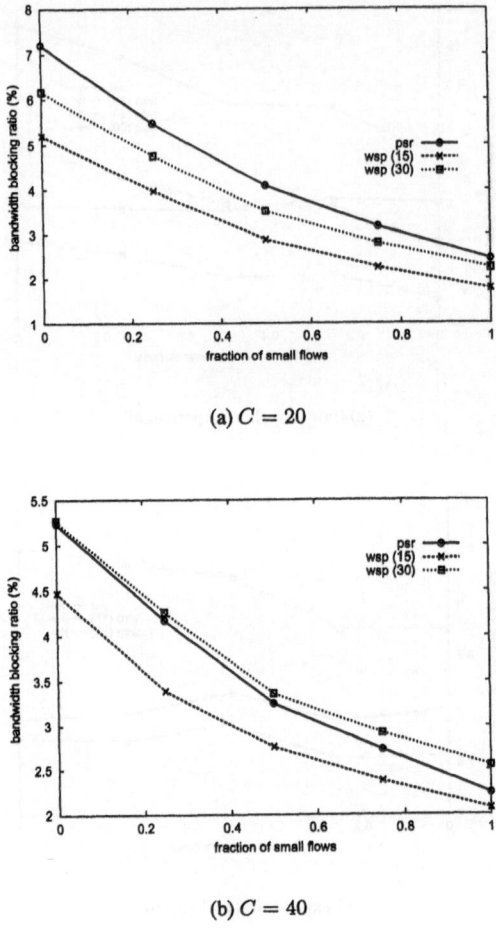

(a) $C = 20$

(b) $C = 40$

Figure 5.12.   Variable bandwidth traffic

nario with short and long flows. An important thing to note is that in all the scenarios the performance of *psr* is insensitive to the holding times of flows.

The behavior of *psr* is not surprising since Erlang formula is known to be applicable even when the flow holding times are not exponentially distributed and blocking probability depends only on the load, i.e., the ratio of arrival rate and service rate. For the above case of two types of flows, the aggregate arrival rate, λ, is given by $\lambda = \lambda_1 + \lambda_2$ and the mean holding time, $1/\mu$, is given by $\frac{1}{\mu} = \frac{1}{\mu_1} \frac{\lambda_1}{\lambda_1 + \lambda_2} + \frac{1}{\mu_2} \frac{\lambda_2}{\lambda_1 + \lambda_2}$. This heterogeneous traffic can then be treated as equivalent to homogeneous traffic with arrival rate λ, mean holding time $1/\mu$ and the corresponding load $\lambda/\mu = \lambda_1/\mu_1 + \lambda_2/\mu_2$. So for a given load, the blocking probability would be the same irrespective of the mean holding times of individual flows. That is why the performance of the theoretical scheme, *vcr* depends only on the overall offered load and not on the types of traffic. The practical scheme, *psr* also behaves similarly and hence *psr* can be employed *as is* to route flows with mixed holding times.

Varying bandwidth requests

Now, consider the case of traffic with 2 types of flows, each requesting for *different amount of bandwidth* but having the same mean holding time. The bandwidth requests of flows are derived uniformly from a range: 0.5 to 1.5 for *small* flows and 1.5 to 2.5 for *large* flows, i.e., the mean bandwidth of small flows is 1 while it is 2 for large flows. The holding times of all the flows are drawn from an exponential distribution with mean 60 sec. The performance is measured varying the mix of small and large flows. Figure 5.12(a) shows the bandwidth blocking ratio as a function of the fraction of small flows. First thing to note is that *psr* performs poorly when the majority of flows are large. However, as the number of small flows increases, it approaches the performance of *wsp*(30). The reason is that routing under *psr* is independent of the amount of bandwidth requested while *wsp* is conscious of the bandwidth requested. However, when the link capacity is much larger than a flow's bandwidth request, *psr* performs fine even though it is unconscious of the requested amount. To illustrate this, we increased the capacity of all links to 40 and measured the performance of both the schemes under similar load conditions as the previous case. Figure 5.12(b) shows that *psr* performs as well as *wsp*(30) when all the flows are large and approaches *wsp*(15) as the number of small flows increases. In the following, we argue further that when bandwidth requests are significantly smaller than the link capacity, it is not necessary for *psr* to differentiate between different bandwidth requests.

In [56], it was shown that when the capacity of a link is large, the blocking probability of a flow of type i can be approximated as follows. Suppose that type i flow requests for d_i units of bandwidth and the load of type i flows on link l is ν_l^i. The blocking probability for type i flows on link l is given by $b_l^i = \frac{d_i}{\delta} E(\frac{\sum \nu_l^i d_i}{\delta}, \frac{c_l}{\delta})$, where δ is an "equivalent rate" given by $\delta = \frac{\sum \nu_l^i d_i^2}{\sum \nu_l^i d_i}$. In other words, the ratio of blocking probabilities of flow types i and j would be same as the ratio of their bandwidth requests, i.e., $\frac{b_i}{b_j} \approx \frac{d_i}{d_j}$. This implies that $\frac{\lambda_1 b_1}{\lambda_2 b_2} = \frac{\phi_1}{\phi_2}$, i.e., the blocking rate of flows of a type is proportional to their fraction in the total offered load. Consequently, performance of a equalization based proportional routing scheme would be same with or without categorizing the flows into different classes. However, *psr* has to be extended to route flows with relatively large bandwidth requests, since it is possible that a path that is good for one bandwidth request may not be even feasible for another

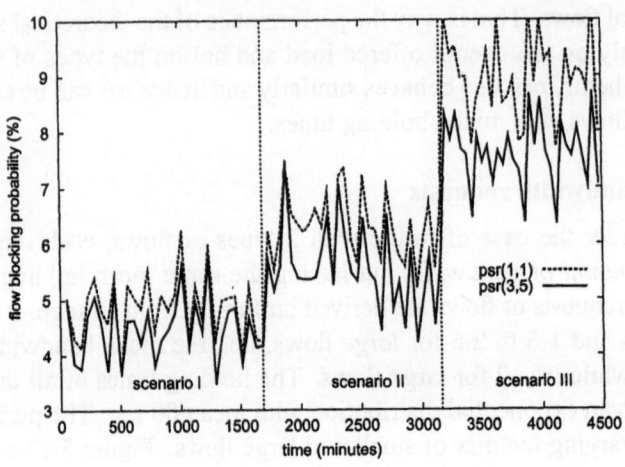

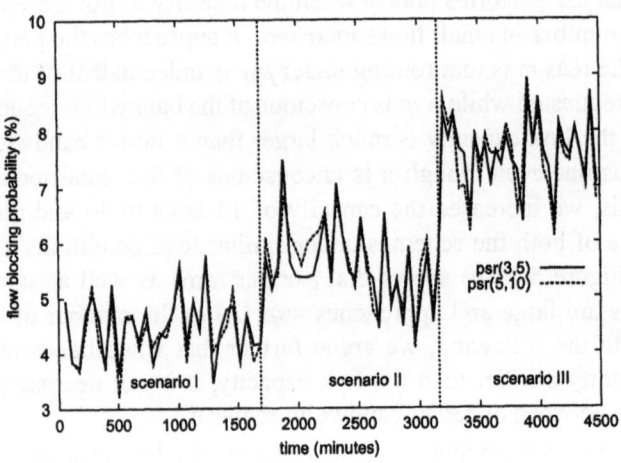

Figure 5.13. Performance of *psr* under different $(\eta,\hat{\gamma})$ settings: (a) (3,5) *vs* (1,1); (b) (3,5) *vs* (5,10).

bandwidth request. In such a case, since the amount of bandwidth requested by a flow is known at the time of path selection, it makes sense to utilize this knowledge in categorizing them into bandwidth classes and routing them accordingly. Considering that in practice link capacities are much larger than an individual flow's bandwidth request, *psr* can be used *as is* to route heterogeneous traffic in most cases.

2.5   Sensitivity of *psr*

We now study the sensitivity of *psr* to the settings of its configurable parameters, η and $\hat{\gamma}$. These parameters control the observation period between successive computations of proportions. While η specifies the number of cycles in an observation period, $\hat{\gamma}$ gives the number of blocks permitted per path in a cycle and thus indirectly controls the length of a cycle. We have experimented with several settings of $(\eta,\hat{\gamma})$ and here we present the results of three different settings: $(1,1)$, $(3,5)$, and $(5,10)$ in Figure 5.13. Two separate graphs are shown for readability. The traffic patterns and loads are varied to see the adaptivity of *psr* under different settings. In scenario I, a load of 0.35 is offered between border nodes and in scenario II, an additional load of 0.05 is offered between hot pairs only and this hot load is increased to 0.10 in scenario III. Under all settings, *psr* adapts quickly to traffic scenario changes. But $psr(3,5)$ blocks lesser flows than $psr(1,1)$ while no discernible difference between $psr(3,5)$ and $psr(5,10)$. The performance difference between $psr(1,1)$ and $psr(3,5)$ is more evident in scenario III where the overall offered load is high. In general, fewer the blocks permitted in a cycle, lesser the effect of proportional routing. Relatively longer cycles are needed to get a good estimate of right proportions. Also, from the perspective of stability it is better to change proportions gradually to reduce oscillations. From these results, we observe that 3 cycles and 5 blocks per path per cycle seem to work fine and beyond that *psr* is relatively insensitive to its parameter settings.

2.6   Routing Stability

An essential feature of a good routing scheme is its ability to avoid routing oscillations and thus ensure stability. It was shown [61] that out-of-date information due to larger update intervals can cause route flapping in schemes such as *wsp*. When the utilization on a link is low, an update causes all the source nodes to prefer routes along this path, resulting in a rapid increase in its utilization. Similarly when the utilization is high, an update causes all the sources to shun this link and consequently its utilization decreases as the existing flows depart. This synchronization problem is inherent in any global information exchange based QoS routing schemes such as *wsp*. On the other hand, the *psr* scheme doesn't exhibit such route flapping behavior. There are two fundamental reasons for the stability of *psr*. First, in *psr* each source performs routing based on its own *local view* of the network state. Routing based on such a "customized view" avoids the undesirable *synchronized mass reaction* that is inherent in QoS routing scheme based on a global view. Second, *psr* does proportional routing with a proportion assigned to a path reflecting its quality. A relatively better path is favored by sending larger proportion of traffic to it. It doesn't pick just one "best" path. The psr can also cause higher fluctuation occasionally at the end of a cycle due to making some paths ineligible and routing all the load along one or a few eligible paths. However, as proportions stabilize, duration of such fluctuations tend be smaller. Considering all this we claim that a localized proportional routing scheme such as *psr* is intrinsically more stable than a global best-path routing scheme such as *wsp*.

Routing Overhead

We now take a close look at the amount of overhead involved in these two routing schemes. This overhead can be categorized into path selection overhead and information collection overhead. We discuss these two separately in the following.

The *wsp* scheme selects a path by first pruning the links with insufficient available bandwidth and then performing a variant of Dijkstra's algorithm on the resulting graph to find the shortest path with maximum bottleneck bandwidth. This takes at least $O(E \log N)$ time where N is the number of nodes and E is the total number of links in the network. Assuming precomputation of a set of paths R to each destination to avoid searching the whole graph for path selection, it still need to traverse all the links of these precomputed paths. This amounts to an overhead of $O(L)$, where L is the total number of links in the set R. On the other hand, the path selection in *psr* is simply an invocation of *wrrps* whose worst case complexity is $O(|R|)$ which is much less than $O(L)$ for *wsp*.

Now consider the information collection overhead. In *wsp*, each source acquires a network-wide view on the status of links through link state updates. Every router is responsible for maintaining QoS state and generating updates about all the links adjacent to it. These updates are sent either periodically or after a significant change in the resource availability since the last update. They are propagated to all the routers in the network through flooding. As in OSPF [43] each router is responsible for maintaining a consistent QoS state database. This incurs both communication and processing overhead. In contrast, the routers employing *psr* scheme do not exchange any such updates and thus completely do away with this overhead. Only source routers need to keep track of route level statistics and recompute proportions after every observation period. Statistics collection in *psr* involves only increment and decrement operations costing only constant time per flow. The proportion computation procedure in *psr* itself is extremely simple and costs no more than $O(|R|)$.

3.   Approximation of *ebp*

In the previous section, we presented *psr* scheme that approximates *ebr* strategy and shown that it is a viable alternative to global schemes such as *wsp*. In this section, we describe an approximation of *ebp* strategy that is suitable for practical implementation. We refer to this approximation also as *ebp* scheme.

In this section, we first describe the proportion computation procedure in *ebp* that at the end of an observation period computes new proportions based on the offered load and the observed blocking probabilities in that period. We then compare the performance of *ebp* with *psr*, *wsp* and other schemes, and show that *ebp* performs the best among localized proportional routing schemes. Finally, we study the sensitivity and optimality of *ebp* scheme.

3.1 Proportion Computation

The *ebp* strategy can be implemented using the following procedure to compute new proportions after an observation period. First, the current average blocking probability $\bar{b} = \sum_{i=1}^{k} \alpha_{r_i} b_{r_i}$ is computed. Then, the proportion of load onto a path r_i is decreased if its current blocking probability b_{r_i} is higher than the average $\bar{b}$ and increased if b_{r_i} is lower than $\bar{b}$. The magnitude of change is determined based on the relative distance of b_{r_i} from $\bar{b}$ and two configurable parameters to ensure that change is gradual. These parameters are *maximum proportional change*, δ and the corresponding *expected proportional change in blocking probability* ϕ. These parameters capture the relative change in blocking probability corresponding to a change in the load, i.e., if the load is changed by a fraction δ, the blocking probability would change by a fraction ϕ. Based on these parameters, if $b_{r_i} > \bar{b}$, the load onto a path r_i is decreased, i.e.,

$$ \nu_{r_i}^* = \frac{\nu_i}{1 + min(\delta, \frac{b_{r_i} - \bar{b}}{b_{r_i}} \frac{\delta}{\phi})} $$

It is increased if $b_{r_i} < \bar{b}$, i.e.,

$$ \nu_{r_i}^* = \nu_i^* (1 + min(\delta, \frac{\bar{b} - b_{r_i}}{\bar{b}} \frac{\delta}{\phi})) $$

The corresponding proportions would then be $\alpha_i^* = \frac{\nu_{r_i}^*}{\sum_{j=1}^{k} \nu_j^*}$. The *mean time between proportion computations* is controlled by a configurable parameter θ. The blocking performance of the candidate paths are observed for a period θ and at the end of the period the proportions are recomputed. This period θ should be large enough to allow for a reasonable measurement of the quality of the candidate paths and small enough to ensure adaptivity to changing traffic conditions.

3.2 Performance Evaluation

We now evaluate the performance of *ebp* using simulations. The simulation setting used for this study is same as the one presented in Section 4.2.4.0. Only thing we need to mention here is that all the results presented here correspond to simulations on *bigisp* topology. The parameters in the simulation are set

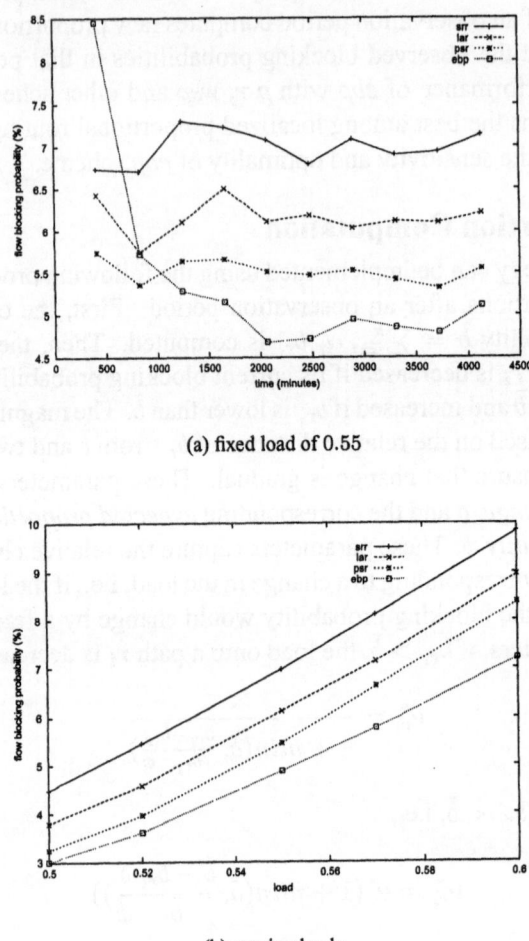

(a) fixed load of 0.55

(b) varying loads

Figure 5.14.   Performance of various localized routing schemes

as follows by default. Any change from these settings is explicitly mentioned wherever necessary. The values for configurable parameters in *ebp* are $\delta = 0.2$, $\phi = 1.0$, $\theta = 60$ minutes (here after written as just *m*). For each pair σ, all the paths between them whose length is at most one hop more than the minimum number of hops (both minhop and minhop+1 paths) are chosen as candidate paths.

We first compare the performance of various localized routing schemes. The *srr* scheme is the sticky random routing scheme described in Section 3.2.1. The *lar* scheme is the learning automata based routing scheme presented in Section 3.2.2. The configurable parameters in *lar* are fixed at $\epsilon = 0.01$ and

$a = 0.02$. Both these schemes perform worse when minhop+1 paths are also made candidates and so we use their performance with minhop paths only as candidates for comparison. The parameters in *psr* scheme are set to $\eta = 3$, $\hat{\gamma} = 5$, and $\psi = 0.8$.

Figure 5.14(a) shows the performance of these schemes when the offered load is fixed at 0.55. The overall blocking probability is shown as a function of time. First thing to note is the convergence of *ebp*. Starting with arbitrary proportions, it gradually adapts and reaches a stable state. All the other schemes converge more quickly but block many more flows than *ebp*. Among these schemes *srr* performs worse. This is because sticky routing directs the whole load onto one path till it blocks, while the other path is relatively idle. Though this scheme is supposed [15] to equalize the blocking rates of candidate paths, it does not ensure that the corresponding proportions are maintained in small time intervals. On the other hand, the *psr* scheme attempts to maintain these proportions while using sticky routing principle to obtain the proportions. Consequently *psr* scheme performs better than *srr*. The *ebp* scheme performs even better than *psr*, particularly when the load is high as as shown in Figure 5.14(b). This figure shows the blocking performance of these schemes under various loads. It can be seen that across all loads the performance of *ebp* scheme is better than the rest of the schemes. Hence we choose *ebp* scheme for localized adaptive proportioning of flows and further study its behavior.

We now compare the performance of *ebp* with *wsp*. Once again, the Figure 5.15(a) shows the overall blocking probability as a function of time with the load set to 0.55. The performance of *wsp* is shown for three different update intervals (in minutes): $0.0, 0.5, 1.0$. The update interval of 0.0 implies that an update is generated for every change in a link's available bandwidth. This corresponds to the performance of a server that keeps track of the precise current state of the network and performs path selection and admission control in a centralized manner. As observed earlier, the impact of update interval on *wsp* is drastic as can be seen from the jump in the blocking probability when the update interval is changed from an unrealistic setting of 0 to a more realistic value of 0.5 minutes or 30 seconds. This shows that best-path routing is good when the bandwidth availability information is accurate which requires frequent updates. On the other hand, our *ebp* scheme without any global updates performs as well as *wsp* with update interval of 30 seconds. This is true across various loads as shown in Figure 5.15(b).

We now study the sensitivity of *ebp* to the values chosen for δ, ϕ, and θ. We first look at the impact of observation interval on the performance of *ebp* by varying the interval from 30 to 90 minutes. The values of δ and *phi* are set to 0.1 and 0.5 and the load is fixed at 0.55. The corresponding results are shown in Figure 5.16(a). With smaller observation interval, *ebp* converges relatively faster. But the overall performance is almost same for all observation

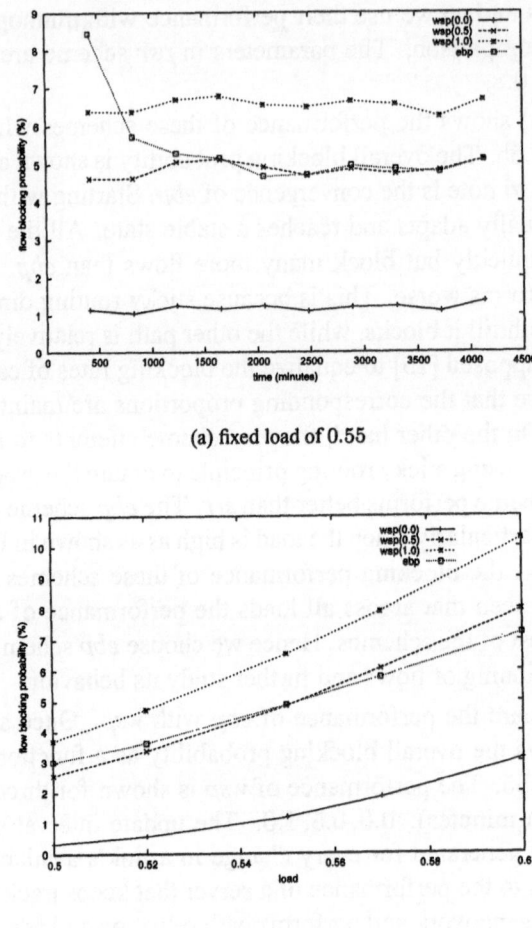

(a) fixed load of 0.55

(b) varying loads

Figure 5.15.   Performance of *ebp* vs *wsp*

interval settings. Figure 5.16(b) shows the results of our simulations when the observation interval θ is fixed at 60 minutes, and δ and *phi* are varied. Here there is some impact of various settings on the performance of *ebp*. However, the difference in overall blocking probabilities is not significant and we can say that the performance of *ebp* is relatively insensitive to its parameter settings.

One important observation that can be made from Figure 5.16 is the significant gap between the performance of *ebp* and the offline optimal proportioning scheme *opr*. While *ebp* is supposed to be a near-optimal strategy, these results show that it may not work as well in practice as in theory. There are two reasons for this gap. First, due to statistical variations the blocking probability

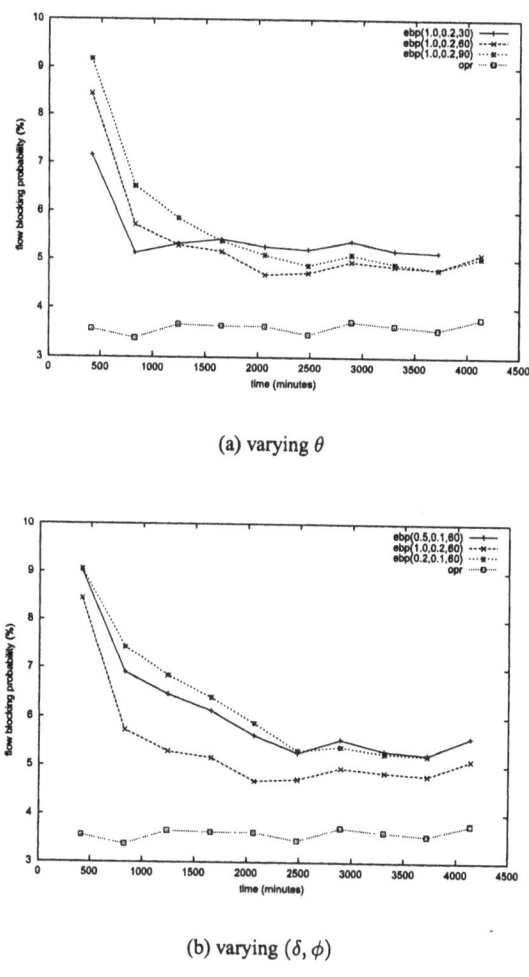

(a) varying θ

(b) varying (δ, ϕ)

Figure 5.16.   Sensitivity of *ebp* scheme

information of a path gathered by a source depends on the observation interval and may not be precise. Second, while in theory it is possible to probe the
quality of a path by routing infinitesimally small proportion of traffic, in reality
non-negligible proportion of traffic has to be sent along each candidate path to
probe its quality. Consequently, bad candidate paths affect the performance of
any practical proportional routing scheme. Hence the performance of localized
schemes such as *ebp* depends critically on the choice of candidate paths. In the
next chapter, we address this problem of how to select a few good candidate
paths.

Chapter 6

CANDIDATE PATH SELECTION

The localized approach to proportional routing described in the previous chapters is simple and has several important advantages. However it has a limitation that routing is done based solely on the information collected locally. A network node under localized QoS routing approach can judge the quality of paths/links only by routing some traffic along them. It would have no knowledge about the state of the rest of the network. While the proportions for paths are adjusted to reflect the changing qualities of paths, the candidate path set itself remains static. To ensure that the localized scheme adapts to varying network conditions, many feasible paths have to be made candidates. It is not possible to preselect a few good candidate paths statically. Hence it is desirable to *supplement localized proportional routing* with a mechanism that *dynamically selects a few good candidate paths*.

Two key questions that arise in candidate path selection are how many paths are needed and how to find these paths. Clearly, the number and the quality of the paths selected as candidates dictate the performance of a proportional routing scheme. There are several reasons why it is desirable to minimize the number of paths used for routing. First, there is a significant overhead associated with establishing, maintaining and tearing down of paths. Second, the complexity of the scheme that distributes traffic among multiple paths increases considerably as the number of paths increases. Third, there could be a limit on the number of explicitly routed paths such as label switched paths in MPLS [57] that can be setup between a pair of nodes. Therefore it is desirable to use *as few paths as possible* while at the same time *minimize the congestion* in the network.

1. Hybrid Approach to QoS Routing

For judicious selection of paths, some knowledge regarding the (global) network state is crucial. This knowledge about resource availability at network nodes, for example, can be obtained through (periodic) information exchange among routers in a network. Because network resource availability changes with each flow arrival and departure, maintaining *accurate* view of network QoS state requires *frequent* information exchanges among the network nodes and introduces both communication and processing overheads. However, these updates would not cause significant burden on the network as long as their frequency is not more than what is needed to convey connectivity information in traditional routing protocols like OSPF [43]. The QoS state of each link could then be piggybacked along with the conventional link state updates. Hence it is important to devise multipath routing schemes that *work well even when the updates are infrequent.*

We propose such a scheme *widest disjoint paths* (wdp) that uses proportional routing — the traffic is proportioned among a few widest disjoint paths. It uses *infrequently* exchanged *global* information for selecting a few good paths based on their long term available bandwidths. It proportions traffic among the selected paths using *local* information to cushion the short term variations in their available bandwidths. Thus the *hybrid* approach to QoS routing adapts at different time scales to the changing network conditions. The rest of the chapter discusses what type of global information is exchanged and how it is used to select a few good paths. It also describes what information is collected locally and how traffic is proportioned adaptively. We first briefly mention some of the related work.

Related Work

Several multipath routing schemes have been proposed for balancing the load across the network. The Equal Cost Multipath (ECMP) [43] and Optimized Multipath (OMP) [64, 65] schemes perform packet level forwarding decisions. ECMP splits the traffic *equally* among multiple equal cost paths. However, these paths are determined statically and may not reflect the congestion state of the network. Furthermore, it is desirable to apportion the traffic according to the quality of each path. OMP is similar in spirit to our work. It also uses updates to gather link loading information, selects a set of best paths and distributes traffic among them. However, our scheme makes routing decisions at the flow level and consequently the objectives and procedures are different.

Another approach to path selection is to precompute maximally disjoint paths [53] and attempt them in some order. This is static and overly conservative. What matters is not the sharing itself but *the sharing of bottleneck links,* which change with network conditions. In our scheme we dynamically select paths such that they are disjoint *w.r.t* bottleneck links.

2. Widest Disjoint Paths

In this section, we present the candidate path selection procedure used in *wdp*. To help determine whether a path is good and whether to include it in the candidate path set, we define *width* of a path and introduce the notion of *width* of a *set of paths*. The candidate path set R_σ for a pair σ is changed only if it increases the width of the set R_σ or decreases the size of the set R_σ without reducing its width. The widths of paths are computed based on link state updates that carry *average residual bandwidth* information about each link. The traffic is then proportioned among the candidate paths using *ebp*.

A basic question that needs to be addressed by any path selection procedure is what is a "good" path. In general, a path can be categorized as good if its inclusion in the candidate path set decreases the overall blocking probability considerably. It is possible to judge the utility of a path by measuring the performance with and without using the path. However, it is not practical to conduct such inclusion-exclusion experiment for each feasible path. Moreover, each source has to independently perform such trials without being directly aware of the actions of other sources which are only indirectly reflected in the state of the links. Hence each source has to try out paths that are likely to decrease blocking and make such decisions with some local objective that leads the system towards a global optimum.

When identifying a set of candidate paths, another issue that requires attention is the sharing of links between paths. A set of paths that are good *individually* may not perform as well as expected *collectively*. This is due to the sharing of *bottleneck* links. When two candidate paths of a pair share a bottleneck link, it may be possible to remove one of the paths and shift all its load to the other path without increasing the blocking probability. Thus by ensuring that candidate paths of a pair do not share bottleneck links, we can reduce the number of candidate paths without increasing the blocking probability. A simple guideline to enforce this could be that the candidate paths of a pair be mutually disjoint, i.e., they do not share *any* links. This is overly restrictive, since even with shared links, some paths can cause reduction in blocking if those links are not congested. What matters is not the sharing itself but *the sharing of bottleneck links*. While the sharing of links among the paths is *static* information independent of traffic, identifying bottleneck links is *dynamic* since the congestion in the network depends on the offered traffic and routing patterns. Therefore it is essential that candidate paths be *mutually disjoint w.r.t bottleneck links*.

To judge the quality of a path, we define *width* of a path as the the residual bandwidth on its bottleneck link. Let $\hat{c}_l$ be the maximum capacity of link l

```
1.    PROCEDURE WIDTH(R)
2.        W = 0
3.        While R ≠ ∅
4.            w* = max_{r∈R} w_r
5.            R* = {r : r ∈ R, w_r = w*}
6.            d* = min_{r∈R*} d_r
7.            r* = {r : r ∈ R*, d_r = d*}
8.            W = W + w*
9.            For each l in r*
10.               c_l = c_l - w*
11.           R = R \ r*
12.       Return W
13.   END PROCEDURE
```

Figure 6.1.   The procedure to compute width for a path set R

and ν_l be the average load on it. The difference $c_l = \hat{c}_l - \nu_l$ is the average residual bandwidth on link l. Then the *width* w_r of a path r is given by $w_r = \min_{l \in r} c_l$. The larger its width is, the better the path is, and the higher its potential is to decrease blocking. Similarly we define *distance* [34] of a path r as $\sum_{l \in r} \frac{1}{c_l}$. The shorter the distance is, the better the path is. The widths and distances of paths can be computed given the residual bandwidth information about each link in the network. This information can be obtained through periodic link state updates. To discount short term fluctuations, the *average residual bandwidth* information is exchanged. Let τ be the update interval and u_l^t be the utilization of link l during the period $(t - \tau, t)$. Then the average residual bandwidth at time t, $c_l^t = (1 - u_l^t)\hat{c}_l$. Hereafter without the superscript, c_l refers to the most recently updated value of the average residual bandwidth of link l.

To aid in path selection, we also introduce the notion of *width* for a *set of paths* R, which is computed as follows. We first pick the path r^* with the largest width w_{r^*}. If there are multiple such paths, we choose the one with the shortest distance d_{r^*}. We then decrease the residual bandwidth on all its links by an amount w_{r^*}. This effectively makes the residual bandwidth on its bottleneck link to be 0. We remove the path r^* from the set R and then select a path with the next largest width based on the just updated residual bandwidths. Note that this change in residual bandwidths of links is local and only for the purpose computing the width of R. This process is repeated till the set R becomes empty. The sum of all the widths of paths computed thus is defined as the *width of R*. Note that when two paths share a bottleneck link, the width of two paths together is same as the width of a single path. The width of a path set computed thus, essentially accounts for the sharing of links between paths.

The procedure to compute the width of a path set R is shown in Figure 6.1. In each iteration, a subset of paths R^* with the largest width w^* are identified (lines 4-5). From these widest paths, a path r^* with the shortest distance d^* is selected (lines 6-7). The width w^* of path r^* is added to the total width W (line 8). The residual capacities of all the links along the path r^* is reduced by an amount w^* (lines 9-10). This in turn affects the widths of other paths in R. The path r^* is removed from the set (line 11) and this process is repeated till the set R becomes empty (line 3). The resulting W is considered to be the width of R. The narrowest path, i.e., the last path removed from the set R is referred to as NARROWEST(R).

Based on this notion of width of a path set, we propose a path selection procedure that *adds* a new candidate path only if its inclusion *increases the width*. It *deletes* an existing candidate path if its exclusion *does not decrease* the total width. In other words, each modification to the candidate path set either *improves the width* or *reduces the number* of candidate paths. The selection procedure is shown in Figure 6.2. First, the load contributed by each existing candidate path is deducted from the corresponding links (lines 2-4). After this adjustment, the residual bandwidth c_l on each link l reflects the load offered on l by all source destination pairs other than σ. Given these adjusted residual bandwidths, the candidate path set R_σ is modified as follows.

The benefit of inclusion of a feasible path r is determined based on the number of existing candidate paths (lines 6-8). If this number is below the specified limit η, the resulting width W_r is the width of $R_\sigma \cup r$. Otherwise, it is the width of $R_\sigma \cup r \setminus$ NARROWEST$(R_\sigma \cup r)$, i.e., the width after excluding the narrowest path among $R_\sigma \cup r$. Let W^+ be the largest width that can be obtained by adding a feasible path (line 9). This width W^+ is compared with width of the current set of candidate paths. A feasible path is made a candidate if its inclusion in set R_σ increases the width by a fraction ψ (line 10). Here $\psi > 0$ is a configurable parameter to ensure that each addition improves the width by a significant amount. It is possible that many feasible paths may cause the width to be increased to W^+. Among such paths, the path r^+ with the shortest distance is chosen for inclusion (lines 11-13). Let r^- be the narrowest path in the set $R_\sigma \cup r$ (line 14). The path r^- is replaced with r^+ if either the number of paths already reached the limit or the path r^- does not contribute to the width (lines 15-16). Otherwise the path r^+ is simply added to the set of candidate paths (lines 17-18). When no new path is added, an existing candidate path is deleted from the set if it does not change the width (lines 20-22). In all other cases, the candidate path set remains unaffected. It is obvious that this procedure always either increases the width or decreases the number of candidate paths.

It should be noted that though *wdp* uses link state updates it does not suffer from the *synchronization* problem unlike global QoS routing schemes such as *wsp*. There are several reasons contributing to the stability of *wdp*: 1) The information exchanged about a link is its *average* not *instantaneous* residual

```
1.    PROCEDURE SELECT(σ)
2.      For each path r in R_σ
3.        For each link l in r
4.          c_l = c_l + (1 − b_r)ν_r
5.      If |R_σ| < η
6.        W_r = WIDTH(R_σ ∪ r), ∀r ∈ R̂_σ \ R_σ
7.      Else
8.        W_r = WIDTH(R_σ ∪ r \ NARROWEST(R_σ ∪ r)), ∀r ∈ R̂_σ \ R_σ
9.      W⁺ = max_{r∈R̂_σ\R_σ} W_r
10.     If (W⁺ > (1 + ψ) WIDTH(R_σ))
11.       R⁺ = {r : r ∈ R̂_σ \ R_σ, W_r = W⁺}
12.       d⁺ = min_{r∈R⁺} d_r
13.       r⁺ = {r : r ∈ R⁺, d_r = d⁺}
14.       r⁻ = NARROWEST(R_σ ∪ r)
15.       If (|R_σ| = η or WIDTH(R_σ ∪ r⁺ \ r⁻) = W⁺)
16.         R_σ = R_σ ∪ r⁺ \ r⁻
17.       Else
18.         R_σ = R_σ ∪ r⁺
19.     Else
20.       r⁻ = NARROWEST(R_σ)
21.       If WIDTH(R_σ \ r⁻) = WIDTH(R_σ)
22.         R_σ = R_σ \ r⁻
23.   END PROCEDURE
```

Figure 6.2.   The candidate path set selection procedure for pair σ

bandwidth and hence less variable; 2) The traffic is proportioned among few "good" paths instead of loading the "best" path based on inaccurate information; 3) Each pair uses only a few candidate paths and makes only incremental changes to the candidate path set; 4) The new candidate paths are selected for a pair only after deducting the load contributed by the current candidate paths from their links. Due to such adjustment even with link state updates, the view of the network for each node would be different; 5) When network is in a stable state of convergence, the information carried in link state updates would not become outdated and consequently each node would have reasonably accurate view of the network. Essentially the nature of information exchanged and the manner in which it is utilized work in a mutually beneficial fashion and lead the system towards a stable optimal state.

We now illustrate how *wdp* scheme selects candidate paths using a simple example. Consider a topology shown in Figure 6.3. Suppose that source *s* has to recompute candidate paths to destination *d*. There are five possible paths between *s* and *d*. Let us assume that *s* is currently using paths via 2 → 5 and 3 → 7, and proportioning traffic equally between them. Further, assume that the average amount of load successfully routed between *s* and *d* is 40. Let the

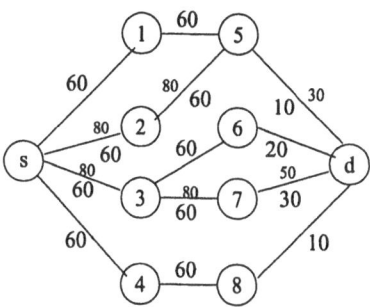

Figure 6.3.   Illustration of *wdp* procedure

average available bandwidths of links, received by source *s* through global link state updates, be as shown in bigger font.

Before recomputing candidate paths, source *s* has to perform local adjustment to discount the bandwidth usage by itself. The source *s* is currently contributing a load of 20 each to paths $2 \rightarrow 5$ and $3 \rightarrow 7$. So the average available bandwidths of links along these paths are correspondingly increased by 20. The new values after local adjustment are shown in smaller font. Essentially the source *s* views the available bandwidth on link $5 \rightarrow d$ as 30 instead of 10 while other sources view it differently.

Now if the maximum number of candidate paths allowed, η, is only two, the candidate path set remains same. This is because the current candidate paths are wider than other paths and replacing any of these paths does not increase the total width. If η is more than two, the path via $3 \rightarrow 6$ is added to the candidate set. Only 4 paths with combined width of 110 would be made candidates even if there is no constraint on the number of candidate paths. The path via $1 \rightarrow 5$ would never be added since it would not increase the total width. Note that though paths $s \rightarrow 3 \rightarrow 6 \rightarrow d$ and $s \rightarrow 3 \rightarrow 7 \rightarrow d$ share a link $s \rightarrow 3$, both are preferred as candidates, since the common link is not the bottleneck. On the other hand, $s \rightarrow 1 \rightarrow 5 \rightarrow d$ is included and $s \rightarrow 2 \rightarrow 5 \rightarrow d$ is excluded since they share a bottleneck link $5 \rightarrow d$. Thus *wdp* selects widest paths that are mutually disjoint w.r.t. bottleneck links.

3.   Performance Analysis

In this section, we evaluate the performance of the proposed hybrid scheme *wdp*. We start with the description of the simulation environment. First, we compare the performance of *wdp* with that of the optimal scheme *opr* and show that *wdp* converges to near-optimal proportions. Furthermore, we demonstrate that the performance of *wdp* is relatively insensitive to the values chosen for the configurable parameters. We then contrast the performance of *wdp* with

η :	maximum number of paths allowed between a pair
ψ :	width increase threshold for changing the path set
τ :	mean time between link state updates
θ :	mean time between computation of proportions
ξ :	mean time between computation of candidate paths

Figure 6.4. Configurable parameters in *wdp*

global QoS routing scheme *wsp* in terms of the overall blocking probability and routing overhead.

3.1   Simulation Environment

The simulation setting used for this study is same as the one presented in Section 4.2.4.0. Once again, the results here correspond to simulations on *bigisp* topology. The values for configurable parameters in *wdp* are set to $\psi = 0.2$, $\tau = 30$ *m*, $\theta = 60$ *m*, $\xi = 180$ *m*. The description of these parameters is given in Figure 6.4. For each pair σ, all the paths between them whose length is at most one hop more than the minimum number of hops are included in the feasible path set $\hat{R}_\sigma$. The amount of offered load on the network ρ is set to 0.55.

3.2   Performance of *wdp*

In this section, we compare the performance of *wdp* and *opr* to show that *wdp* converges to near-optimal proportions using only a few paths for routing traffic. We also demonstrate that *wdp* is relatively insensitive to the settings for the configurable parameters.

Convergence

Figure 6.5 illustrates the convergence process of *wdp*. The results are shown for different values of $\eta = 1 \cdots 4$. Figure 6.5(a) compares the performance of *wdp*, *opr* and *ebp*. The performance is measured in terms of the overall flow blocking probability, which is defined as the ratio of the total number of blocks to the total number of flow arrivals. The overall blocking probability is plotted as a function of time. In the case of *opr*, the algorithm is run offline to find the optimal proportions given the set of feasible paths and the offered load between each pair of nodes. The resulting proportions are then used in simulation for statically proportioning the traffic among the set of feasible paths. The *ebp* scheme refers to the localized scheme used in isolation for adaptively proportioning across all the feasible paths. As noted earlier all paths of length either minhop or minhop+1 are chosen as the set of feasible paths in our study.

There are several conclusions that can be drawn from Figure 6.5(a). First, the *wdp* scheme converges for all values of η. Given that the time between

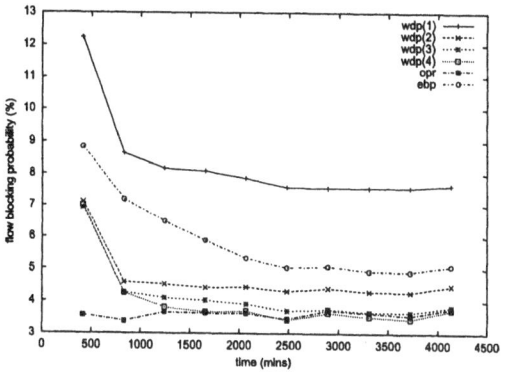

(a) blocking probability

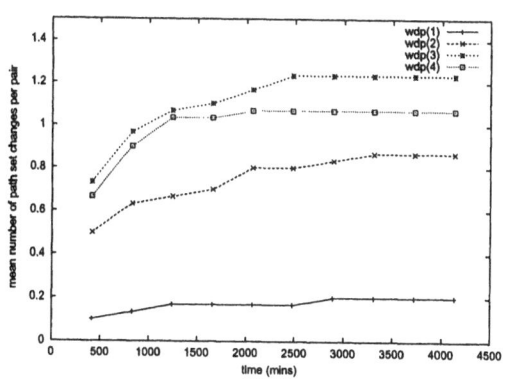

(b) number of changes to candidate paths

Figure 6.5. Convergence process of *wdp*

changes to candidate path sets, ξ, is 180 m, it reaches steady state within (on average) 5 path recomputations per pair. Second, there is a marked reduction in the blocking probability when the number of paths allowed, η, is changed from 1 to 2. It is evident that there is quite a significant gain in using multipath routing instead of single path routing. When the limit η is increased from 2 to 3 the improvement in blocking is somewhat less but significant. Note that in our topology there are at most two paths between a pair that do not share any links. But there could be more than two paths that are mutually disjoint w.r.t bottleneck links. The performance difference between η values of 2 and 3 is an indication that we only need to ensure that candidate paths do not share congested links. However using more than 3 paths per pair helps very little

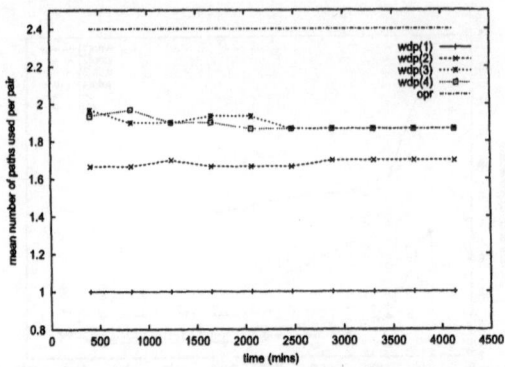

Figure 6.6.   Number of paths used for routing

in decreasing the blocking probability. Third, the *ebp* scheme also converges, albeit slowly. Though it performs much better than *wdp* with single path, it is worse than *wdp* with $\eta = 2$. But when *ebp* is used in conjunction with path selection under *wdp* it converges quickly to lower blocking probability using only a few paths. Finally, using at most 3 paths per pair, the *wdp* scheme approaches the performance of optimal proportional routing scheme.

Figure 6.5(b) establishes the convergence of *wdp*. It shows the average number of changes to the candidate path set as a function of time. Here the change refers to either addition, deletion or replacement operation on the candidate path set R_σ of any pair σ. Note that the cumulative number of changes are plotted as a function of time and hence a plateau implies that there is no change to any of the path sets. It can be seen that the path sets change incrementally initially and after a while they stabilize. Thereafter each pair sticks to the set of chosen paths. It should be noted that starting with at most 3 minhop paths as candidates and making as few as 1.2 changes to the set of candidate paths, the *wdp* scheme achieves almost optimal performance.

We now compare the average number of paths used by a source-destination pair for routing as shown in Figure 6.6. Note that in *wdp* scheme η specifies only the maximum allowed number of paths per pair. The actual number of paths selected for routing depends on their widths. The average number of paths used by *wdp* for η of 2 and 3 are 1.7 and 1.9 respectively. The number of paths used stays same even for higher values of η. The *ebp* scheme uses all the given feasible paths for routing. It can measure the quality of a path only by routing some traffic along that path. The average number of feasible paths chosen are 5.6. In case of *opr* we count only those paths that are assigned a proportion of at least 0.10 by the optimal offline algorithm. The average number of such paths under *opr* scheme are 2.4. These results support our

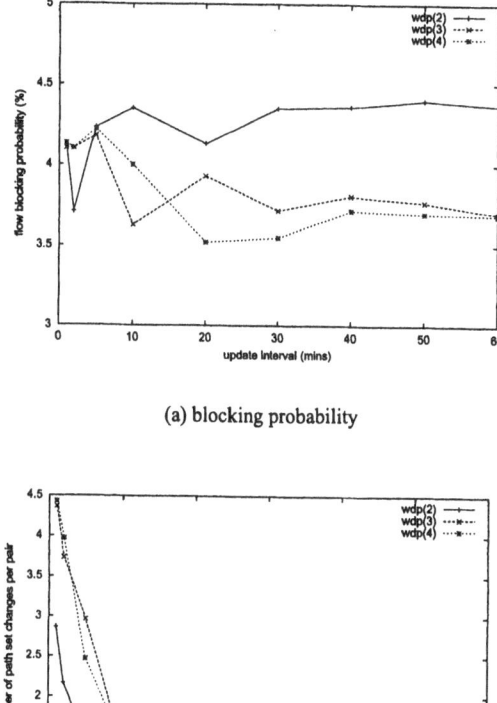

(a) blocking probability

(b) number of path set changes

Figure 6.7. Sensitivity of *wdp* to update interval τ

claim that *ebp* based proportioning over widest disjoint paths performs almost like optimal proportioning scheme while using fewer paths.

Sensitivity

The *wdp* scheme requires periodic updates to obtain global link state information and to perform path selection. To study the impact of update interval on the performance of *wdp*, we conducted several simulations with different update intervals ranging from 1 *m* to 60 *m*. The Figure 6.7(a) shows the flow blocking probability as a function of update interval. At smaller update intervals there is some variation in the blocking probability, but much less variation at larger update intervals. It is also clear that increasing the update interval does not cause any significant change in the blocking probability. To study the effect of update interval on the stability of *wdp*, we plotted the average number of path set changes as a function of update interval in Figure 6.7(b). It shows that the candidate path set of a pair changes often when the updates are frequent. When the update interval is small, the average residual bandwidths of links resemble their instantaneous values, thus highly varying. Due to such variations, paths may appear wider or narrower than they actually are, resulting in unnecessary changes to candidate paths. However, this does not have a significant impact on the blocking performance due to adaptive proportional routing among the selected paths. For the purpose of reducing overhead and increasing stability, we suggest that the update interval τ be reasonably large, while ensuring that it is much smaller than the path recomputation interval ξ.

We also studied the sensitivity of *wdp* to the width increase threshold parameter for changing the path set ψ. The simulations were run for five different values of ψ from 0.10 to 0.30. Figure 6.8(a) shows the flow blocking probability as as a function of ψ. It can be seen that the blocking performance of *wdp* is relatively insensitive to the value of ψ. Once again in Figure 6.8(b) the number of path set changes is shown as a function of ψ. As expected, there are fewer changes as the ψ value increases. However this does not effect the blocking significantly since much wider paths are included in the set anyway. Only those paths that may contribute to the width by a small amount are excluded by choosing higher ψ value. To ensure that good paths are not excluded and not so good paths are not included we suggest setting ψ to a value around 0.20.

3.3 Comparison of *wsp* and *wdp*

We now compare the performance of hybrid QoS routing scheme *wdp* with a global QoS routing scheme *wsp*. The *wsp* is a well-studied scheme that selects the widest shortest path for each flow based on the global network view obtained through link state updates. The information carried in these updates is the residual bandwidth at the instant of the update. Note that *wdp* also employs link state updates but the information exchanged is average residual bandwidth over a period not its instantaneous value. We use *wsp* as a representative of global QoS routing schemes as it was shown to perform the best among similar schemes such as shortest widest path (*swp*), shortest distance path (*sdp*). In the following, we first compare the performance of *wdp* with *wsp* in terms of flow blocking probability and then the routing overhead.

Blocking Probability

Figure 6.9(a) shows the blocking probability as a function of update interval τ used in *wsp*. The τ for *wdp* is fixed at 30 m. The offered load on the network

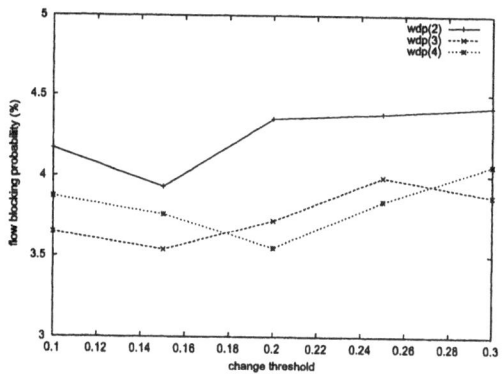

(a) blocking probability

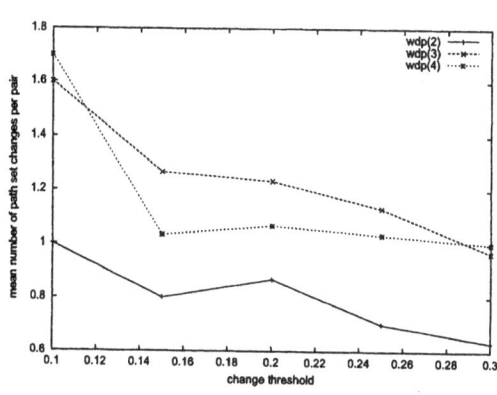

(b) number of path set changes

Figure 6.8. Sensitivity of *wdp* to change threshold ψ

ρ was set to 0.55. It is clear that the performance of *wsp* degrades drastically as the update interval increases. The *wdp* scheme, using at most two paths per pair and infrequent updates with $\tau = 30\ m$, blocks fewer flows than *wsp*, that uses many more paths and frequent updates with $\tau = 0.5\ m$. The performance of *wdp* even with a single path is comparable to *wsp* with $\tau = 1.5\ m$. Figure 6.9(b) displays the flow blocking probability as a function of offered network load ρ which is varied from 0.50 to 0.60. Once again, the τ for *wdp* is set to 30 m and the performance of *wsp* is plotted for 3 different settings of τ: 0.5, 1.0 and 2.0 m. It can be seen that across all loads the performance of *wdp* with $\eta = 2$ is better than *wsp* with $\tau = 0.5$. Similarly with just one path, *wdp* performs better than *wsp* with $\tau = 2.0$ and approaches the performance of $\tau = 1.0$ as the load

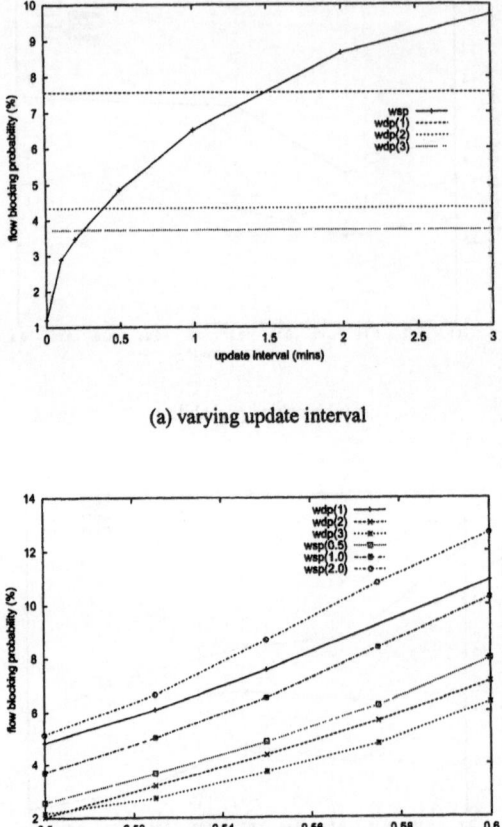

(a) varying update interval

(b) varying load

Figure 6.9.   Performance comparison of *wdp* and *wsp*

increases. It is also worth noting that *wdp* with two paths rejects significantly fewer flows than with just one path, justifying the need for multipath routing.

It is interesting to observe that even with a single path and very infrequent updates *wdp* outperforms *wsp* with frequent updates. There are several factors contributing to the superior performance of *wdp*. First, it is the nature of information used to capture the link state. The information exchanged about a link is its *average* not *instantaneous* residual bandwidth and hence less variable. Second, before picking the widest disjoint paths, the residual bandwidth on all the links along the current candidate path are adjusted to account for the load offered on that path by this pair. Such a *local adjustment* to the global information makes the network state appear differently to each source. It is as if each source receives a customized update about the state of each link. The sources that are currently routing through a link perceive higher residual bandwidth on that link than other sources. This causes a source to continue using the same path to a destination unless it finds a much wider path. This in turn reduces the variation in link state and consequently the updated information does not get outdated too soon. In contrast, *wsp* exchanges highly varying instantaneous residual bandwidth information and all the sources have the same view of the network. This results in mass synchronization as every source prefers *good* links and avoids *bad* links. This in turn increases the variance in instantaneous residual bandwidth values and causes route oscillation[1]. The *wdp* scheme, on the other hand, by selecting paths using both local and global information and by employing *ebp* based adaptive proportioning delivers stable and robust performance.

Routing Overhead

Now we compare the amount of overhead incurred by *wdp* and *wsp*. This overhead can be categorized into per flow routing overhead and operational overhead. We discuss these two separately in the following.

The *wsp* scheme selects a path by first pruning the links with insufficient available bandwidth and then performing a variant of Dijkstra's algorithm on the resulting graph to find the shortest path with maximum bottleneck bandwidth. This takes at least $O(E \log N)$ time where N is the number of nodes and E is the total number of links in the network. Assuming precomputation of a set of paths R_σ to each destination, to avoid searching the whole graph for path selection, it still needs to traverse all the links of these precomputed paths to identify the widest shortest path. This amounts to an overhead of $O(L_\sigma)$, where L_σ is the total number of links in the set R_σ. On the other hand, in *wdp* one of the candidate paths is chosen in a weighted round robin fashion whose complexity is $O(\eta)$ which is much less than $O(L_\sigma)$ for *wsp*.

Now consider the operational overhead. Both schemes require link state updates to carry residual bandwidth information. However the frequency of updates needed for proper functioning of *wdp* is no more than what is used to carry connectivity information in traditional routing protocols such as OSPF. Therefore, the average residual bandwidth information required by *wdp* can be piggybacked along with the conventional link state updates. Hence, *wdp* does not cause any additional burden on the network. On the other hand, the *wsp* scheme requires frequent updates consuming both network bandwidth and processing power. Furthermore *wsp* uses too many paths. The *wdp* scheme uses only a few preset paths, thus avoiding per flow path setup. Only admission con-

---

[1]Some remedial solutions were proposed in [1, 2] to deal with the inaccuracy at a source node. However, the fundamental problem remains and the observations made here still apply.

trol decision need to be made by routers along the path. The other overheads incurred only by *wdp* are periodic proportion computation and candidate path computation. The proportion computation procedure is extremely simple and costs no more than $O(\eta)$. The candidate path computation amounts to finding η widest paths and hence its worst case time complexity is $O(\eta N^2)$. However, this cost is incurred only once every ξ period. Considering both the blocking performance and the routing cost, we conclude that *wdp* yields much higher throughput with much lower overhead than *wsp*.

Chapter 7

HIERARCHICAL PROPORTIONAL ROUTING

We have shown that proportional routing schemes alleviate congestion in the network by distributing load among multiple "good" paths instead of overloading a single "best" path. However these schemes assume that each router in the network is aware of the topology and the state of the whole network. This is referred to as *flat* routing and under flat routing, each router participates in link state updates and maintains detailed information about the entire network. This introduces significant burden on every router and as the size of the network grows, the overhead at each router increases tremendously. To provide a scalable solution, *hierarchical routing* is suggested [43, 4] as an alternative to flat routing. To make the proportional routing schemes scale well to large networks we need to extend them to provide hierarchical routing.

1. Hierarchical Routing

Under hierarchical routing, a network is divided into multiple areas.[1] The routing within the area is flat with each router having detailed information about routers and links in that area. But the routers have only sketchy *aggregate* information about other areas. To route traffic destined for other areas, a source router may select a partial higher level path, based on the aggregate information, that gets expanded, based on the detailed information, at the ingress border router of each area along the path. Such a hierarchical routing reduces the overhead at each router by limiting the scope of link state updates and maintaining only summary information about other areas. Examples of hierarchical routing are inter-area routing in OSPF [43] and ATM Forum's PNNI [4].

---

[1] Also referred to as peer groups in PNNI [4].

The hierarchical routing approach while reduces the burden on a router, introduces inaccuracy in the information available for routing. Hence the performance of a hierarchical routing scheme depends heavily on how information about an area is aggregated and how it is utilized in routing across areas. For providing QoS routing across areas, in addition to *topology aggregation*, *QoS state aggregation* must also be performed. Topology aggregation is concerned with capturing the structure of an area which is relatively static. There are several proposals for topology aggregation such as [32, 31]. QoS state aggregation is concerned with summarizing the QoS state of an area which is quite dynamic. It is somewhat straightforward to summarize the the QoS state when best-path routing is employed. The state of routing between a pair of routers is given by the state of the best path. The best-path based hierarchical routing is studied in [19]. On the other hand, *it is not obvious how to aggregate the topology and state when multiple paths are used to route traffic between a pair of routers*. This chapter addresses the issue of aggregation under multipath proportional routing, and the selection of paths based on the aggregated information.

We propose an aggregation method that summarizes the state of multiple paths between two routers using a single metric. This metric in essence captures the traffic carrying capacity of multiple paths between a pair of routers. Our approach is somewhat similar in spirit to the proposal in [41] but the actual path selection schemes are quite different. We propose two inter-area routing schemes based on this aggregate metric: *hierarchical widest disjoint paths* (hwdp) and *hierarchical widest border routers* (hwbr). The *hwdp* scheme is a hierarchical source routing scheme where a source router selects a set of higher level skeletal paths to the destination as candidates and proportions flows among them. The *hwbr* scheme is a hierarchical next-hop routing scheme that selects only the next-hop border routers which in turn select higher level paths to the destination. Both these schemes use our *widest disjoint paths* (wdp), a flat multipath proportional routing scheme described in the previous chapter, for intra-area routing to expand the skeletal higher level paths to actual physical paths. They essentially differ in the way the network outside an area is aggregated by a border router and propagated to the interior routers. The following sections describe the proposed aggregation metric and the hierarchical multipath proportional routing schemes based on this metric. We also evaluate the performance of the proposed schemes and compare them with best-path based hierarchical routing schemes discussed in [19].

2.   Topology and State Aggregation

Topology aggregation is concerned with capturing the structure of an area which is relatively static. There are several proposals for topology aggregation such as [31, 32]. We take a simple approach where a border router makes other

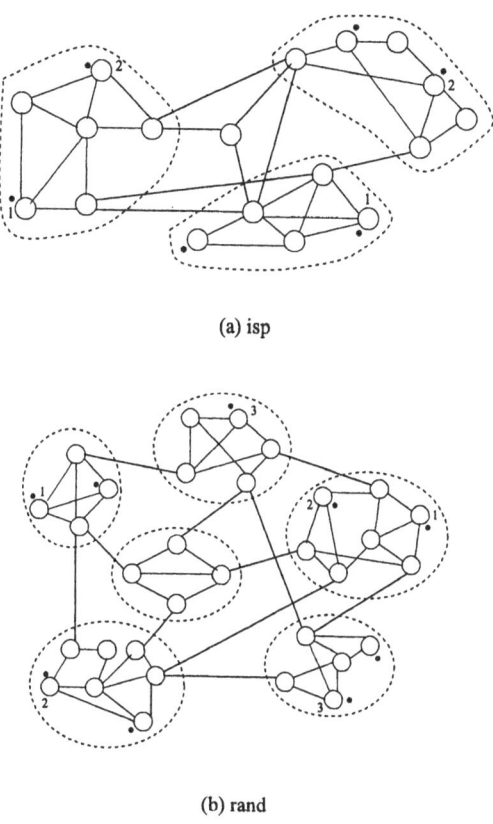

(a) isp

(b) rand

Figure 7.1. Topologies used in our study

routers in its area appear as directly connected to it by a *logical link*. This approach is similar to the one employed by OSPF [43]. Then state aggregation is about summarizing the state of the network by assigning the attributes to these logical links. When best-path routing is used to route traffic within an area, it is straightforward to perform state aggregation: the attributes of a logical link are that of the best-path. For example, when traffic within an area is routed along shortest paths, then the *distance* of the logical link between a pair of routers would simply be the distance of the shortest path. On the other hand, it is not obvious how to summarize the state when multiple paths are used to route traffic between a pair of routers.

The multipath proportional routing scheme *wdp* described in the previous chapter lends itself very well to aggregation. Note that *wdp* selects candidate paths such that the total *width* of these candidates is as large as possible. This width essentially captures the traffic-carrying capacity of all the paths between

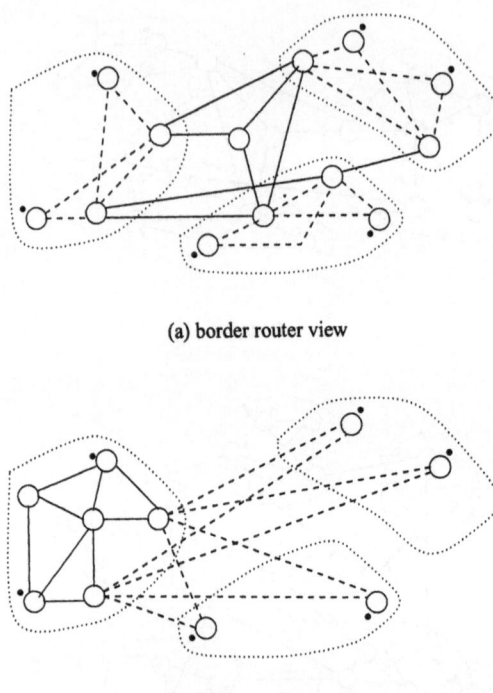

(a) border router view

(b) source router view

Figure 7.2.   Different views of the isp topology

a pair of routers. Hence, we propose to summarize the state of multiple paths between a pair of routers by a single metric, *the width of its candidate paths*. This metric not only provides more accurate but also more stable description of the state of the network than the best-path metric which could change quite frequently. Given a set of candidate paths R_σ of a pair σ, its width, W_σ can be computed using the procedure shown in Figure 6.1, i.e., $W_\sigma = \text{WIDTH}(R_\sigma)$.

We propose two hierarchical routing schemes based on this aggregate metric. These schemes are similar in that both employ *wdp* for intra-area routing and exchange aggregate width information among border routers. However, they differ in the amount and thus granularity of aggregate information propagated to interior routers by border routers and consequently in the selection of partial higher level paths to destinations.

3.   Hierarchical Source Routing

Suppose that a border router injects all the information it receives from other border routers into its area. Then an interior router also would have same level of aggregate information as any border router. For routing to a destination in a different area, an interior source router could then select an higher level skeletal path consisting of border routers. Each border router along the skeletal path would then use intra-area routing to find a path to the next border router along the path till the destination is reached. This is referred to as *hierarchical source routing*.

The *wdp* scheme can be extended naturally to provide hierarchical source routing. At the higher level each area is represented by a set of logical links. The average available bandwidth of a logical link corresponding to a pair σ is set to the width of σ, W_σ. Now we can form a higher level network consisting of physical backbone links and the logical links representing each area. For example, consider the isp topology shown in Figure 7.1(a). The corresponding higher level view of border routers is shown in Figure 7.2(a). Under hierarchical source routing, an interior source router would also have a view similar to border routers since border routers propagate their view as is to interior routers also.

Given this higher level view of the network, a source router can apply *wdp* scheme shown in Figure 6.2 *as is* on this higher level network to identify a set of widest disjoint paths as candidates and perform *ebp* based proportioning among these higher level candidate paths. However, there is one difference in that two paths that are considered disjoint at the higher level may not be really disjoint at the lower level. We can be conservative and treat two higher level paths as disjoint only if they do not pass through the same area. In our study, enforcement of such a constraint did not make much difference since such paths were anyway not chosen due to the sharing of inter-area links. We refer to this hierarchical version of *wdp* where *wdp* is used for both intra-area and inter-area routing as *hwdp*.

4. Hierarchical Next-hop Routing

It is likely that a border router would provide much less detailed information about the network to an interior router than what is available to itself. It is desirable that a border router further summarize the information it has before propagating it to the interior routers. This reduces the communication overhead on the network and avoids the information overload at interior routers. In other words, a border router not only provides summarized view of its area to other border routers but also presents aggregated view of the network outside the area to its interior routers. For example, the view of a source router corresponding to the isp topology in Figure 7.1(a) is shown in Figure 7.2(b) while the border routers view it as shown in Figure 7.2(a). Under such a realistic scenario, an interior router may have only sufficient information to select an exit border router to reach a destination. It could first select a border router, and then identify an internal path to the border router. The border router which has more detailed information about the backbone network finds a path to a border router in the destination area which would in turn route to the destination. We refer to this approach of selecting the higher level next-hop instead of complete higher level path as *hierarchical next-hop routing*.

We propose a hierarchical next-hop routing scheme which is referred to as *hierarchical widest border routers* (hwbr). Under *hwbr* scheme, a border router further aggregates multiple higher level logical paths to a destination into a single metric, width of candidate logical paths. It then passes this per destination width information to interior routers. Let $R_{x,d}$ be the set of higher level candidate paths from a border router x to a destination d. Then the border router x would pass the aggregate width $W_{x,d} = \text{WIDTH}(R_{x,d})$ for each destination d to its interior routers. An interior router selects a border router as its next-hop to a destination based on the width from the border router to the destination and the width of candidate paths from it to the border router.

There is a tradeoff in selecting a border router that has better external route to a destination and a border router to which there is a better internal route. However, what matters from the perspective of reducing blocking probability is the width of bottleneck segment. So we define $width(s, d, x)$ from a source s to a destination d via border router x as follows. Let $W_{s,x}$ be the width from source router s to border router x and $W_{x,d}$ be the width from x to destination d. Then $width(s, d, x) = \min(W_{s,x}, W_{x,d})$. Given the widths of all border routers to a destination, a border router with the largest width is chosen as the next-hop to the destination. Note that the next-hop selection is not per flow but periodically recomputed after performing local width adjustment as is the case with selection of candidate paths in *wdp* and *hwdp*. When more than one border router is allowed to be a candidate next-hop, then they are selected in the order of their widths and flows are proportioned among them using the *ebp* strategy.

5.   Performance Evaluation

In this section, we evaluate the performance of the proposed hierarchical multipath proportional routing schemes. These schemes are compared with the flat *wdp* scheme. We also compare them against the hierarchical best-path routing schemes. We choose *wsp* as a representative of best-path routing approach as it is shown to perform the best among such schemes [1, 19]. We refer to the hierarchical version of *wsp* as *hwsp*. Under *hwsp*, the state of routing between a pair of routers is summarized by the widest shortest path. Hence, the bandwidth and the hop count of a logical link between a pair of routers is given by bottleneck bandwidth and the hop count of the corresponding widest shortest path. The *hwsp* is a hierarchical source routing scheme where a source router selects a widest shortest higher level logical path based on the bandwidths and hop counts of the logical links. This skeletal path is then expanded by the border routers using *wsp* to route within the area. We now describe the simulation environment.

5.1   Simulation Environment

Figures 7.1(a) and 7.1(b) show the topologies used in our study. The *isp* topology is a slightly altered version of the topology used in our earlier study and also by others [1, 34]. The *rand* topology is similar to the one used in [19]. In both the topologies the *thin* links connect routers within an area and the *thick* links are the backbone links. All thin links are assumed to have same capacity of 20 units and all thick links 30 units. Flows arriving into the network are assumed to require one unit of bandwidth. Hence a link with capacity C can accommodate at most C flows simultaneously. The flow dynamics of the network are modeled as follows. The routers labeled with a *dot* are considered to be source or destination routers. Flows arrive into the system according to a Poisson process with rate λ. A pair of source and destination routers are chosen randomly from the set of all such pairs. The holding time of a flow is exponentially distributed with mean $1/\mu$. The overall load offered on the network is then $\rho = \lambda/\mu$. We set the average holding time of a flow, $1/\mu$ to 1 minute and vary the arrival rate λ depending upon the desired overall load ρ.

The default values for the configurable parameters of the schemes being simulated are as follows. The update interval in *wsp* and *hwsp* is set to 0.5 minutes or 30 seconds and in the rest of our proportional routing schemes it is set to 30 minutes. The observation interval between recomputations of proportions in the flat *wdp* scheme is set to 40 minutes and candidate path selection is done every 120 minutes. Same settings are used for inter-area routing in both *hwdp* and *hwbr* and the corresponding values for intra-area routing are 20 minutes and 60 minutes respectively. These values are chosen such that inter-area paths are changed more gradually than intra-area paths for the purpose of stability. The number of candidate paths allowed between a pair of nodes is set to 3 for flat routing and intra-area routing.

5.2   Convergence and Adaptivity

Figure 7.3 illustrates the convergence and adaptivity of multipath proportional routing schemes. The results corresponding to isp topology are shown in Figure 7.3(a) and that of rand topology are shown in Figure 7.3(b). The performance is measured in terms of the overall flow blocking probability, which is defined as the ratio of the number of blocked flows to the total number of flow arrivals. The overall flow blocking probability is plotted as a function of time. We consider two traffic scenarios. In scenario I, all source-destination pairs are offered an equal amount of load and in scenario II, certain "hot pairs" exchange more traffic than other pairs. We have chosen two hot pairs marked

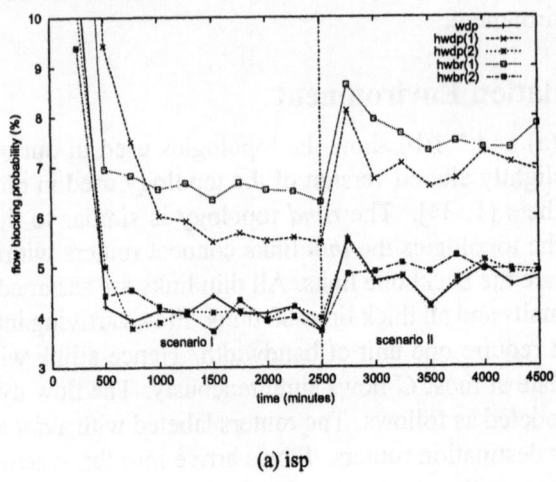

(a) isp

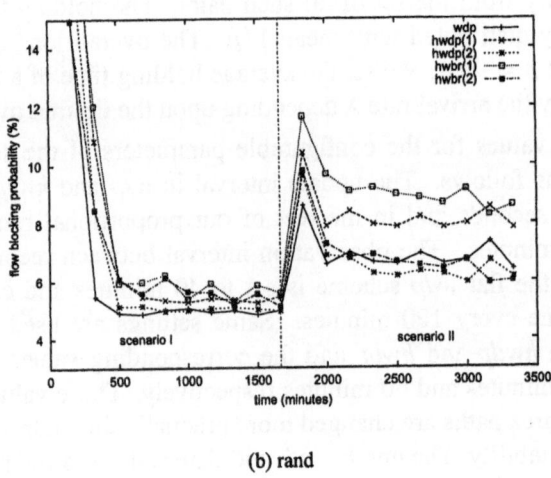

(b) rand

Figure 7.3. Convergence and adaptivity of proportional routing schemes

by 1 and 2 in case of isp. Similarly in case of rand three pairs are chosen as hot. In case of isp, we offer a load of 200 in scenario I which is equally split between all source-destination pairs and in scenario II a load of 150 that is equally split among all pairs and a load 50 that is equally split among hot pairs only. The corresponding values for rand case are 300, 200 and 100 respectively.

The performance of *hwdp* is shown for two cases with the number of higher level candidate paths set to 1 and 2. Similarly two plots correspond to *hwbr* with the number of candidate border routers set to 1 and 2. We start the simulations with scenario I and switch to scenario II after sometime. There are several observations that can be made from these results. Starting with an arbitrary set of candidates and proportions, all the multipath proportional routing schemes gradually converge to stable state. This gets disturbed when the traffic scenario changes and the blocking probability shoots up as the candidates and their proportions chosen for one traffic pattern are not perfect for a different traffic pattern. Consequently the proposed proportional routing schemes gradually adapt to the new traffic conditions and converge to stable state again.

With a single candidate path *hwdp* performs better than *hwbr* with a single candidate border router. This is because in *hwbr*, due to lack of sufficient information, many sources may select the same border router as next-hop to many destinations. On the other hand, *hwdp* avoids this problem by using more detailed information in the selection of skeletal path to the destination. However, when the number of candidates is increased to two, there is almost no difference in blocking performance between the source routing scheme *hwdp* and the next-hop routing scheme *hwbr*. These results also indicate that there is significant gain in using multiple candidates for inter-area routing also. The gain is more pronounced in case of isp than in rand and also in scenario II with hot pairs and thus non-uniform load than in scenario I with uniform load. Finally, both the proposed hierarchical schemes with two candidates perform as well as the flat *wdp* scheme.

5.3   Blocking Performance

Figures 7.4(a) and 7.4(b) show the performance of these schemes under different load conditions for isp and rand topologies respectively. The blocking probability is plotted as a function of offered load which is varied from 180 to 220 in case of isp and 260 to 340 in case of rand. Note that the update interval in *wsp* and *hwsp* is set to 30 seconds while that in all our proportional routing schemes is set to 30 minutes. We also show the performance of flat *wsp* scheme with update interval of 0 as a reference. This corresponds to the case of instantaneous updates, i.e., an update generated for every change in a link's available bandwidth or a server that keeps track of the current state of the network and performs path selection and admission control in a centralized manner.

First, let us look at the impact of update interval and state aggregation on the performance of *wsp* scheme. There is a drastic increase in the blocking probability when the update interval is changed from an unrealistic setting of 0 to a more realistic value of 30 seconds. This shows the impact of inaccuracy on best-path routing schemes and that they work well only with very frequent updates. Essentially, best-path routing is suitable for centralized routing but not appropriate for distributed routing. The hierarchical version of *wsp*, *hwsp* fares worse than *wsp*. This is because the impact of inaccuracy introduced by

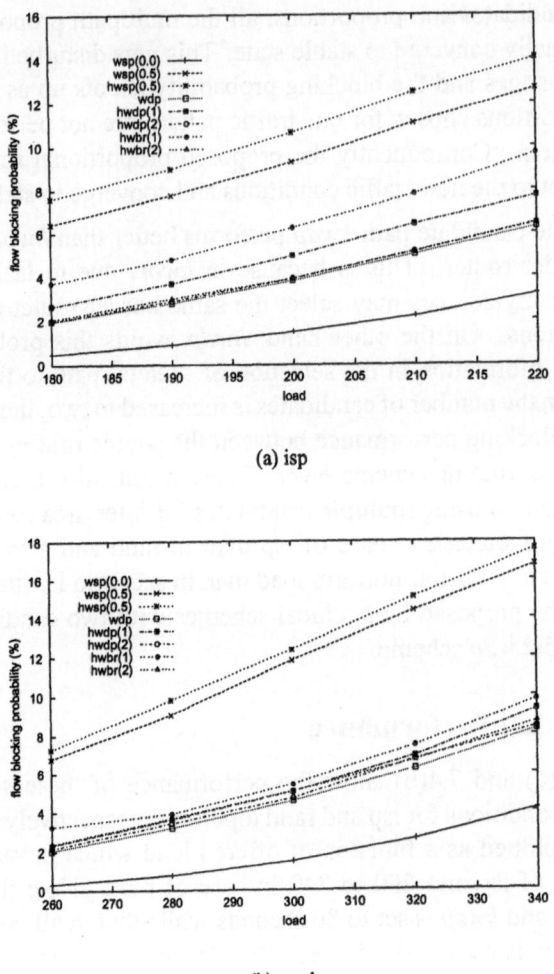

(a) isp

(b) rand

Figure 7.4.   Performance comparison with best-path routing schemes

the best-path based aggregation gets further amplified by the selection of the best higher level logical path based on inaccurate information.

Now, let us compare the proposed hierarchical proportional routing schemes with best-path routing schemes *wsp* and *hwsp*. It is expected that with always up-to-date information (update interval of 0), the flat *wsp* scheme would perform better than proportional routing schemes. However, it is surprising that with only aggregate information and update interval of 30 minutes, hierarchical proportional routing schemes perform much better than even the flat *wsp* scheme with update interval of 30 seconds. It is worth noting that this is the case even when the number of candidates is limited to one in both *hwdp* and *hwbr* schemes. Furthermore, the gap in blocking performance between the proposed schemes and *wsp* widens with the larger rand topology. These results demonstrate that hierarchical multipath proportional routing schemes with their minimal update overhead are more suitable than best-path routing schemes for routing in large networks.

Chapter 8

CONCLUSIONS AND FUTURE WORK

In this monograph, we focused on *proportional routing* approach as an alternative to *best-path routing* approach for providing QoS routing. While best-path routing schemes select the best path for each incoming flow, proportional routing schemes proportion flows among a set of candidate paths. The best-path routing schemes require frequent exchange of network state, imposing both the communication overhead on the network and processing overheads on core routers. On the other hand, a proportional routing scheme can select a few good candidate paths using infrequently exchanged global information and proportion flows among candidates using only locally collected information. We have described one such scheme that selects *widest disjoint paths* as candidates and proportions flows among these paths using a simple local *equalization of blocking probabilities* strategy. We have shown that our proportional routing scheme yields higher throughput with lower overhead than best-path routing schemes.

We have extended our proportional routing approach to provide hierarchical routing across large networks that are divided into multiple areas. We proposed an aggregation method that summarizes the state of multiple paths between two routers in an area using a single metric. We presented two hierarchical multipath routing schemes *hierarchical widest disjoint paths* (hwdp) and *hierarchical widest border routers* (hwbr) that are based on this aggregate metric. We evaluated their performance and shown that the proposed hierarchical multipath routing schemes perform as well as flat multipath routing scheme *wdp*. Furthermore, we demonstrated that these schemes with only aggregate information outperform even the flat best-path routing scheme *wsp* having detailed information about the network.

We have used analytical methods and simulation models to verify the proposed proportional routing schemes and compare them with other schemes.

However it would be ideal if we could implement and validate them in a real setting. But it is not possible to deploy them on a large scale and conduct experiments with an actual network. A reasonable alternative would be to simulate the core network while using FreeBSD or Linux machine implementations of our schemes to act as edge routers. Since the proposed schemes essentially add more intelligence at the edge routers and leave the functionality of core routers untouched, this serves the purpose. We are specifically interested in the amount of complexity introduced by these schemes at the edge routers. This experimental setup would help measure the storage requirements and processing speeds of these schemes. The real traffic conditions can be simulated by running actual traffic traces.

This monograph assumes a service model where admission of flows is controlled such that every admitted flow receives the expected quality of service. While such a service is very much desirable, there is an immediate need for providing what is referred to as *better best-effort service*. Under such a service, there are neither admission controls nor per-flow guarantees. Instead, a better service is provided to all the traffic by balancing the load across the network and by utilizing the resources efficiently. This service based on multipath routing is likely to be more readily adopted by network service providers as an alternative to the existing shortest path routing based best-effort service. Provision of this *better best-effort service* entails tackling many of the issues that are addressed in this monograph in the context of QoS routing: How to select a few good candidate paths and how to split traffic among them. The ideas presented in this monograph such as local adjustment to global updates and selection of widest paths that are mutually disjoint w.r.t. bottleneck links as candidates are very much applicable in this setting too. One key difference is that due to lack of admission control, a source would not be able to gather information such as blocking probability of flows locally. So, what requires to be addressed is how to proportion traffic (in the absence of local information) among a set of candidate paths selected by *wdp*. This and other related problems in deploying *better best-effort service* needs to be investigated.

References

[1] G. Apostolopoulos, R. Guerin, S. Kamat, S. Tripathi, "Quality of Service Based Routing: A Performance Perspective", ACM SIGCOMM 1998.

[2] G. Apostolopoulos, R. Guerin, S. Kamat, S. Tripathi, "Improving QoS Routing Performance under Inaccurate Link State Information," ITC'16, pp. 1351-1362, June 1999.

[3] G. Ash, *Dynamic Routing in Telecommunications Networks*, McGraw-Hill, 1998.

[4] ATM Forum, Private Network-Network Interface, Specification Version 1 (PNNI 1.0), March 1996.

[5] D. Awduche, J. Malcolm, J. Agogbua, M. O'Dell, J. McManus, "Requirements for Traffic Engineering over MPLS," RFC-2702, September 1999.

[6] R.Callon, P. Doolan, N. Feldman, A. Fredette, G. Swallow, and A. Viswanathan, "A Framework for Multiprotocol Label Switching," IETF Internet Draft, Work in Progress, September 1999.

[7] S. Chen and K. Nahrstedt, "An Overview of Quality-of-Service Routing for the Next Generation High-Speed Networks: Problems and Solutions," IEEE Network Magazine, Special Issue on Transmission and Distribution of Digital Video, 1998.

[8] S. Chen and K. Nahrstedt, "Distributed Quality-of-Service Routing in High-Speed Networks Based on Selective Probing," LCN'98, October 1998.

[9] J. Chen, P. Druschel, and D. Subramanian, "A New Approach to Routing with Dynamic Metrics", IEEE INFOCOM 1999.

[10] T. Coleman and Y. Li, "An Interior, Trust Region Approach for Nonlinear Minimization Subject to Bounds," in *Journal on Optimization, Vol. 6*, pp. 418–445, SIAM, 1996.

[11] E. Crawley, R. Nair, B. Rajagopalan, H. Sandick, "A Framework for QoS-Based Routing in the Internet", RFC 2386, August 1998.

[12] N. G. Duffield, K. K. Ramakrishnan, and A. Reibman, "SAVE: An Algorithm for Smoothed Adaptive Video over Explicit Rate Networks", IEEE INFOCOM'98, San Francisco, CA, March 1998.

[13] R. F. Farmer and I. Kaufman, "On the Numerical Evaluation of Some Basic Traffic Formulae," in *Networks, Vol 8.*, pp. 153–186, John Wiley and Sons, Inc., 1978.

[14] Wu-chi Feng, "Rate-constrained Bandwidth Smoothing for the Delivery of Stored Video", SPIE Multimedia Computing and Networking 1997.

[15] R.J. Gibbens, F.P. Kelly, and P.B. Key, "Dynamic Alternative Routing: Modelling and Behaviour", Teletraffic Science, pp. 1019-1025, Elsevier, Amsterdam, 1989.

[16] M. Grossglauser, S. Keshav and D. Tse, "RCBR : A Simple and Efficient Service for Multiple Time-Scale Traffic", Proc. ACM SIGCOMM, pp. 219-230, Aug 1995.

[17] R. Guerin, S. Kamat, A. Orda, T. Przygienda, D. Williams, "QoS Routing Mechanisms and OSPF Extensions", *Work in Progress*, Internet Draft, March 1997.

[18] R. Guerin, A. Orda, "QoS-Based Routing in Networks with Inaccurate Information: Theory and Algorithms", IEEE Infocom 1997.

[19] F. Hao, and E.W. Zegura, "On Scalable QoS Routing: Performance Evaluation of Topology Aggregation", IEEE INFOCOM 2000. 1997.

[20] A.A. Jagers and E.A. Van Doorn, "On the Continued Erlang Loss Function," Op. Res. Lett., vol. 5, pp. 43-46, 1986.

[21] D.L. Jagerman, "Some properties of the Erlang Loss function," Bell Systems Technical Journal, vol. 53, No. 3, March 1974.

[22] D.L. Jagerman, "Methods in traffic calculation," AT&T Bell Lab Technical Journal, vol. 63, No. 7, September 1984.

[23] J.S. Kaufman, "Blocking in a Shared Resource Environment," IEEE Trans. Commun. vol. COM-29, pp. 1474-1481, 1981.

[24] F.P. Kelly, "Routing in Circuit-Switched Networks: Optimization, Shadow Prices and Decentralization", Advances in Applied Probability 20, 112-144, 1988.

[25] F.P. Kelly, "Routing and capacity Allocation in Networks with Trunk Reservation", Mathematics of Operations Research, 15:771-793, 1990.

[26] F.P. Kelly, "Dynamic Routing in Stochastic Networks", In *Stochastic Networks*, ed. F.P. Kellyand R.J. Williams, Springer-Verlag, 169-186, 1995.

[27] F.P. Kelly, "Loss Networks," The Annals of Applied Probability, vol. 1, pp. 319-378, 1991.

[28] F.P. Kelly, "Fixed Point Models of Loss Networks," J. Austr. Math. Soc., Ser. B, 31, pp. 204-218, 1989.

[29] P.B. Key, and G.A. Cope, "Distributed Dynamic Routing Schemes", IEEE Communucations Magazine, pp. 54-64, Oct 1990.

[30] Murali Kodialam, and T. V. Lakshman, "Minimum Interference Routing with Applications to MPLS Traffic Engineering", INFOCOM 2000.

[31] T. Korkmaz and M. Krunz, "Source-oriented Topology Aggregation with Multiple QoS Parameters in Hierarchical ATM Networks", IWQOS 1999.

[32] W.C. Lee, "Topology Aggregation for Hierarchical Routing in ATM Networks", Computer Communications Review, vol. 25, no. 2, pp. 82-92.

[33] Q. Ma, P. Steenkiste, and H. Zhang, "Routing High Bandwidth Traffic in Max-Min Fair Share Networks," In *Proc. ACM SIGCOMM'96*, October 1996, Stanford, CA.

[34] Q. Ma, P. Steenkiste, "On Path Selection for Traffic with Bandwidth Guarantees", IEEE ICNP 1997.

[35] Q. Ma, P. Steenkiste, "Routing Traffic with Quality-of-Service Guarantees in Integrated Services Networks", NOSSDAV 1998.

[36] Q. Ma, "Quality-of-Service Routing in Integrated Services Networks", Ph.D Dissertation, School of Computer Science, Carnegie Mellon University, January 1998.

[37] S. McCanne, V. Jacobson, and M. Vetterli, "Receiver-driven Layered Multicast," ACM SIGCOMM 1996

[38] J.M. McManus and K.W. Ross, "Video on Demand over ATM: Constant-rate Transmission and Transport", Proc. IEEE INFOCOM, pp. 1357-1362, March 1996.

[39] D. Mitra, and J.B. Seery, "Comparative Evaluations of Randomized and Dynamic Routing Strategies for Circuit-Switched Networks", IEEE Trans. on Communications, vol. 39, no. 1, pp. 102-116, January 1991.

[40] D. Mitra, J.A. Morrison, and K.G. Ramakrishnan, "ATM Network Design and Optimization: A Multirate Loss Network Framework," IEEE/ACM Transactions on Networking, vol. 4, no. 4., pp. 531-543, August 1996.

[41] M. Montgomery and G. de Veciana, "Hierarchical Source Routing through Clouds", IEEE INFOCOM 1998.

[42] J.A. Morrison, K.G. Ramakrishnan, and D. Mitra, "Refined asymptotic approximations to loss probabilities and their sensitivities in shared unbuffered resources," SIAM J. Appl. Math., 1998.

[43] J. Moy, "OSPF Version 2", Request For Comments 2328, Internet Engineering Task Force, April 1998.

[44] K.S. Narendra and P. Mars, "The Use of Learning Algorithms in Telephone Traffic Routing - A Methodlogy", Automatica, vol. 19, no. 5, pp. 495-502, 1983.

[45] K.S. Narendra and M.A.L. Thathachar, "On the Behavior of a Learning Automaton in a Changing Environment with Application to Telephone Traffic Routing", IEEE Trans. on Systems, Man and Cybernetics, vol. SMC-10, no.5, pp. 262-269, May 1980.

[46] S. Nelakuditi, R.P. Tsang, Z.-L. Zhang, "Quality-of-Service Routing without Global Information Exchange", IWQOS 1999.

[47] S. Nelakuditi, Z.-L. Zhang, and R.P. Tsang, "Adaptive Proportional Routing: A Localized QoS Routing Approach", IEEE INFOCOM'00, March 2000.

[48] S. Nelakuditi, Z.-L. Zhang, R.P. Tsang, and D.H.C. Du, "Adaptive Proportional Routing: A Localized QoS Routing Approach", IEEE/ACM Transactions on Networking, Dec 2002.

[49] S. Nelakuditi, S. Varadarajan, and Z-L. Zhang, "On Localized Control in Quality-of-Service Routing". Submitted to IEEE Transactions on Automatic Control, Special Issue on Systems and Control Methods for Communication Networks.

[50] S. Nelakuditi, and Z.-L. Zhang, "On Selection of Paths for Multipath Routing", IWQOS'01, June 2001.

[51] S. Nelakuditi, and Z.-L. Zhang, "Localized Adaptive Proportioning Approach to QoS Routing", IEEE Communications Magazine, June 2002.

[52] S. Nelakuditi and Z.-L. Zhang, "Hierarchical Multipath Routing", Technical Report, April 2002.

[53] N. Taft-Plotkin, B. Bellur, and R. Ogier, "Quality-of-Service Routing using Maximally Disjoint Paths", IWQOS 1999.

[54] M. Powell, "A Fast Algorithm for Nonlinear Constrained Optimization Calculations," in *Numerical Analysis, ed. G.A. Watson, Lecture Notes in Mathematics, Vol 630*, Springer Verlag, 1978.

[55] J.W. Roberts, "Teletraffic Models for the Telecom 1 Integrated Services Network," in Proc. Internet Teletraffic Congress-10, Session 1.1, paper #2.

[56] J. Roberts, U. Mocci, and J. Virtamo, "Broadband Network Teletraffic," LNCS 1155, Springer Verlag, 1996.

[57] E. Rosen, A. Viswanathan, and R. Callon, "Multiprotocol Label Switching Architecture," IETF Internet Draft, Work in Progress, August 1999.

[58] K.W. Ross, "Multiservice Loss Models for Broadband Telecommunication Networks", Springer-Verlag, 1995.

[59] A. Shaikh, J. Rexford, K. Shin, "Evaluating the Overheads of Source-Directed Quality-of-Service Routing", ICNP 1998.

[60] A. Shaikh, J. Rexford, K. Shin, "Efficient Precomputation of Quality-of-Service Routes", NOSSDAV 1998.

[61] A. Shaikh, J. Rexford, K. Shin, "Load-Sensitive Routing of Long-Lived IP Flows", ACM SIGCOMM 1998.

[62] P.R. Srikantakumar and K.S. Narendra, "A Learning Model for Routing in Telephone Networks", SIAM J. Control and Optimization, vol. 20, no. 1, pp. 34-57, January 1982.

[63] O. Verscheure, X. Garcia, G. Karlsson, and J.P. Hubaux, "User-Oriented QoS in Packet Video Delivery", IEEE Network, November/December 1998.

[64] C. Villamizar, "OSPF Optimized Multipath (OSPF-OMP), " Internet Draft, February 1999.

[65] C. Villamizar, "MPLS Optimized Multipath (MPLS-OMP)", Internet Draft, February 1999.

[66] Z. Wang, J. Crowcroft, "Quality-of-Service Routing for Supporting Multimedia Applications", IEEE JSAC Sept 1996

[67] D. Wijesekera and J. Srivastava, "Quality of Service (QoS) Metrics for Continuous Media", Multimedia Tools and Applications, Vol 2, No 3, Sept 1996, pp. 127-166.

[68] E. W. Zegura, K.L. Calvert, and S. Bhattacharjee, "How to Model an Internetwork," IEEE INFOCOM 1996.

[69] Z. Zhang, C. Sanchez, B. Salkewicz, E. Crawley, "Quality of Service Extensions to OSPF", *Work in Progress*, Internet Draft, September 1997.

Index